LECTURES ON SYMPLECTIC MANIFOLDS

Other Monographs in this Series

No. 1. Irving Kaplansky: *Algebraic and analytic aspects of operator algebras*
2. Gilbert Baumslag: *Lecture notes on nilpotent groups*
3. Lawrence Markus: *Lectures in differentiable dynamics*
4. H.S.M. Coxeter: *Twisted honeycombs*
5. George W. Whitehead: *Recent advances in homotopy theory*
6. Walter Rudin: *Lectures on the edge-of-the-wedge theorem*
7. Yozo Matsushima: *Holomorphic vector fields on compact Kähler manifolds*
8. Peter Hilton: *Lectures in homological algebra*
9. I. N. Herstein: *Notes from a ring theory conference*
10. Branko Grünbaum: *Arrangements and spreads*
11. Irving Glicksberg: *Recent results on function algebras*
12. Barbara L. Osofsky: *Homological dimensions of modules*
13. Michael Rabin: *Automata on infinite objects and Church's problem*
14. Sigurdur Helgason: *Analysis on Lie groups and homogeneous spaces*
15. R. G. Douglas: *Banach algebra techniques in the theory of Toeplitz operators*
16. Joseph L. Taylor: *Measure algebras*
17. Louis Nirenberg: *Lectures on linear partial differential equations*
18. Avner Friedman: *Differential games*
19. Béla Sz.-Nagy: *Unitary dilations of Hilbert space operators and related topics*
20. Hyman Bass: *Introduction to some methods of algebraic K-theory*
21. Wilhelm Stoll: *Holomorphic functions of finite order in several complex variables*
22. O. T. O'Meara: *Lectures on linear groups*
23. Mary Ellen Rudin: *Lectures on set theoretic topology*
24. Melvin Hochster: *Topics in the homological theory of modules over commutative rings*
25. Karl. W. Gruenberg: *Relation modules of finite groups*
26. Irving Reiner: *Class groups and Picard groups of group rings and orders*
27. H. Blaine Lawson, Jr.: *The quantitative theory of foliations*
28. T. A. Chapman: *Lectures on Hilbert cube manifolds*

Conference Board of the Mathematical Sciences
REGIONAL CONFERENCE SERIES IN MATHEMATICS

supported by the
National Science Foundation

Number 29

LECTURES ON SYMPLECTIC MANIFOLDS

by

ALAN WEINSTEIN

Published for the
Conference Board of the Mathematical Sciences
by the
American Mathematical Society
Providence, Rhode Island

Expository Lectures
from the CBMS Regional Conference
held at the University of North Carolina
March 8–12, 1976

Prepared by the American Mathematical Society
with partial support from National Science Foundation grant MPS-74-23180.
AMS (MOS) subject classifications (1970). Primary 53C15.

Library of Congress Cataloging in Publication Data

Weinstein, Alan, 1943-
Lectures on symplectic manifolds.

(Regional conference series in mathematics ; no. 29)
"Expository lectures from the CBMS regional conference held at the University of North Carolina, March 8-12, 1976."
Bibliography: p.
1. Symplectic manifolds--Addresses, essays, lectures.
I. Conference Board of the Mathematical Sciences.
II. Title. III. Series.
QA1.R33 no. 29 [QA649] 510'.8s [516'.362] 77-3399
ISBN 0-8219-1679-9

Second printing, with corrections, 1979
Printed in the United States of America

CONTENTS

Introduction 1
Lecture 1. Symplectic manifolds and lagrangian submanifolds, examples 3
Lecture 2. Lagrangian splittings, real and complex polarizations, Kähler manifolds 7
Lecture 3. Reduction, the calculus of canonical relations, intermediate polarizations 11
Lecture 4. Hamiltonian systems and group actions on symplectic manifolds 15
Lecture 5. Normal forms 22
Lecture 6. Lagrangian submanifolds and families of functions 25
Lecture 7. Intersection Theory of lagrangian submanifolds 29
Lecture 8. Quantization on cotangent bundles 31
Lecture 9. Quantization and polarizations 35
Lecture 10. Quantizing lagrangian submanifolds and subspaces, construction of the Maslov bundle 39
References 45

Introduction

I like to think of symplectic geometry as playing the role in mathematics of a language which can facilitate communication between geometry and analysis. On the one hand, since the cotangent bundle of any manifold is a symplectic manifold, many phenomena and constructions of differential topology and geometry have symplectic "interpretations", some of which lead to the consideration of symplectic manifolds other than cotangent bundles. On the other hand, the category of symplectic manifolds has formal similarities to the categories of linear spaces used in analysis. The problem of constructing functorial relations which respect these similarities is one aspect of the so-called *quantization* problem; using the solutions of this problem which are presently available, one can construct analytic objects (e.g., solutions of partial differential equations, representations of groups) from symplectic ones.

The most important objects in symplectic geometry, after the symplectic manifolds themselves, are the so-called *lagrangian* submanifolds. They arise in a multitude of ways in the symplectic interpretation of geometric phenomena; on the other hand, they correspond under quantization relations to elements (or classes of elements) of the linear spaces of analysis.

The oldest symplectic manifold of my acquaintance occurs in Lagrange's later work (1808) on celestial mechanics [**LA 1**]. Lagrange wrote the equations of notion for the orbital elements $(x_1, \ldots, x_6)$ of a planet, under the effect of perturbations, in the form $\partial H/\partial x_i = \Sigma a_{ij}(x) dx_j/dt$, where $a_{ij}(x)$ is a 6 by 6 skew-symmetric matrix, and he showed that a suitable choice of the x-coordinate system puts these equations in the form now known as Hamilton's equations. Indeed, the development of symplectic geometry is closely connected with developments in classical mechanics; nevertheless, since there are several extensive studies available on symplectic mechanics ([**AB**], [**AN 4**], [**AN V**], [**SI**], [**SR**]), I shall largely neglect classical mechanics and dynamical systems in the present lectures. (When physics appears, it will usually be quantum mechanics.) We also recommend to the reader the book [**GU**], which discusses in great detail many topics which are merely mentioned here.

The first six sections of these notes contain a description of some of the basic constructions and results on symplectic manifolds and lagrangian submanifolds. §7, on intersections of lagrangian submanifolds, is still mostly internal to symplectic geometry, but it contains some applications to mechanics and dynamical systems. §§8, 9, and 10 are devoted to various aspects of the quantization problem. In §10 there is a feedback of ideas from quantization theory into symplectic geometry itself.

In preparing these lectures, I have had the benefit of several opportunities to present parts of the material in preliminary form, including two courses at Berkeley, the Nordic Summer School of Mathematics (Grebbestad, 1975), and the seminar Demazure-Laudenbach-Poenaru (Orsay, 1976). I would like to thank the students and colleagues in these audiences for their contributions in the form of suggestions, questions, and puzzled expressions, all of which have helped me to improve my formulation of this material.

The regional conference at which these lectures were given was organized by Pat Eberlein and Robbie Gardner. I would like to thank them for their invitation; in addition, I think that the other participants share my gratitude for their organizational efforts and for their inspired (partly by circumstance) choice of the Quail Roost Conference Center as the meeting site. Notes taken at the conference by Gerald Chachere, John Jacob, and Dean Payne have been of great help to me in the preparation of this manuscript.

Finally, I would like to acknowledge the hospitality of the Institute des Hautes Études Scientifiques, where these lectures were prepared, and the financial support of the National Science Foundation.

Lecture 1. Symplectic Manifolds and Lagrangian Submanifolds, Examples

There are three levels on which one may define the notion of a symplectic structure:

(A) *Algebra.* A symplectic structure on a finite-dimensional real vector space V is an antisymmetric bilinear form Ω on V such that the associated map $\widetilde{\Omega}: V \longrightarrow V^*$ defined by $\widetilde{\Omega}(v)(w) = \Omega(v, w)$ is an isomorphism.

(A/G) *Algebra/Geometry.* A symplectic structure on a [smooth] vector bundle E over a space [manifold] X is a continuous [smooth] family $\Omega = \{\Omega_x\}$ of symplectic structures on the fibres of E. The associated object $\widetilde{\Omega} = \{\widetilde{\Omega}_x\}$ is then a bundle isomorphism from E to E^*.

(G) *Geometry.* A symplectic structure on a smooth manifold P is a symplectic structure Ω in sense (A/G) on the tangent bundle TP which, considered as an exterior differential 2-form on P, is closed, i.e. $d\Omega = 0$.

REMARK. There is a common generalization of (A) and (A/G): one may consider symplectic structures on modules over arbitrary commutative rings (see [N]; except for the ring of complex numbers, we will not use this generalization, nor will we consider symplectic structures on Banach spaces (see [C], [MD], [SW]).

An isomorphism between symplectic objects in any of the three categories is called a *symplectomorphism*, or *canonical transformation.*

Historical note. The word *symplectic* was invented by Hermann Weyl [WE], who substituted Greek for Latin roots in the word *complex* to obtain a term which would describe a group related to line complexes but which would not be confused with complex numbers. (I owe this reference to Souriau, who is himself the inventor of the term *symplectomorphism.*)

Here are some examples of symplectic objects.

In (A), let U be any real vector space. Then $U \oplus U^*$ has a "canonical" symplectic structure Ω_U defined by

$$\Omega_U(u_1 \oplus u_1^*, u_2 \oplus u_2^*) = u_2^*(u_1) - u_1^*(u_2).$$

Any symplectic vector space is symplectomorphic to one of these. (See Lecture 2.)

Analogously, in (A/G), the Whitney sum $E \oplus E^*$, where E is any vector bundle over X, is a symplectic vector bundle. *Not* every symplectic vector bundle is isomorphic to one of these.

In (G), the simplest example of a symplectic manifold is $\mathbf{R}^{2n}$ with the symplectic structure $\Omega = \Sigma_{i=1}^n dx_i \wedge d\xi_i$, where $(x_1, \ldots, x_n, \xi_1, \ldots, \xi_n)$ are the coordinates. Note that $\widetilde{\Omega}(\partial/\partial x_i) = d\xi_i$ and $\widetilde{\Omega}(\partial/\partial\xi_i) = -dx_i$. We shall see in Lecture 5 that every symplectic manifold is *locally* symplectomorphic to this one.

If X is any manifold, its cotangent bundle T^*X, considered as a manifold, carries a canonical symplectic structure Ω_X which generalizes that on $\mathbf{R}^{2n} \approx T^*\mathbf{R}^n$. To define Ω_X, we first consider the 1-form ω_X on T^*X defined by $\omega_X(b) = b \circ T_b\pi$, where $\pi : T^*X \longrightarrow X$ is the natural projection and b is any element of T^*X. A local coordinate system $(x_1, \ldots, x_n)$ on X extends to coordinates $(x_1, \ldots, x_n, \xi_1, \ldots, \xi_n)$ on T^*X. Since $(T\pi)(\partial/\partial x_i) = \partial/\partial x_i$ and $(T\pi)(\partial/\partial\xi_i) = 0$, we have $\omega_X = \Sigma_{i=1}^n \xi_i dx_i$. Now $\Omega_X = -d\omega_X$ is given in the same coordinates by $\Omega_X = \Sigma_{i=1}^n dx_i \wedge d\xi_i$, so it is a symplectic structure. Since the construction of Ω_X is completely canonical, the diffeomorphism $T^*f\colon T^*Y \longrightarrow T^*X$ induced by any diffeomorphism $f\colon X \longrightarrow Y$ is a symplectomorphsim with respect to the structures Ω_X and Ω_Y.

The simplest examples of compact symplectic manifolds are given by the orientable surfaces. In fact, if S is any orientable 2-manifold, any nonvanishing 2-form (volume element) Ω on S is a symplectic structure. The total volume $\int_S \Omega$ is a symplectic invariant of (S, Ω); it is a theorem of Moser **[MO]** that is the only one, if S is fixed. (See Lecture 5.) Greene and Shiohama **[GR]** have recently extended Moser's result to open surfaces of finite area.

In general, if (P, Ω) is a compact symplectic manifold of dimension $2n$, Ω^n is a volume element on P, so the de Rham cohomology class $[\Omega^n] = H^{2n}(P; \mathbf{R})$ is nonzero. Since $[\Omega^n] = [\Omega]^n$, $[\Omega] \in H^2(P; \mathbf{R})$ and all of its powers through the nth must be nonzero as well. The existence of such an element of $H^2(P; \mathbf{R})$ is a necessary condition for the compact manifold P to admit a symplectic structure. Another necessary condition of a homotopy-theoretic nature is the existence of a symplectic structure on the bundle TP. It is not known whether these conditions are sufficient. In case P is noncompact, Gromov **[GRM]** has shown that any symplectic structure on the bundle TP is homotopic to a symplectic structure on P. (The 2-form is actually *exact*.)

Certain subobjects of symplectic objects are particularly important. If W is a subspace of a symplectic vector space (V, Ω), the orthogonal space $W^\perp$ is defined to be

$$\{v \in V \mid \Omega(v, w) = 0 \quad \text{for all} \quad w \in W\}.$$

We have $\dim(W) + \dim(W^\perp) = \dim(V)$. W is called:

isotropic	if $W \subseteq W^\perp$,
coisotropic	if $W \supseteq W^\perp$,
lagrangian	if $W = W^\perp$,
symplectic	if $W \cap W^\perp = 0$.

Note that for W to be isotropic [symplectic] means that the restriction of the bilinear form Ω from V to W is identically zero [a symplectic structure].

The four adjectives listed above are applied in the same way to subbundles of symplectic vector bundles and submanifolds of symplectic manifolds. If Q is a submanifold of P, its tangent bundle TQ may be considered as a subbundle of $T_QP = \bigcup_{q\in Q} T_qP$. Q is called an isotropic [etc.] submanifold of P if TQ is an isotropic [etc.] subbundle of T_QP, i.e. if each T_qQ is an isotropic [etc.] subspace of T_qP.

We also apply one of the four adjectives to an immersion $i: Q \longrightarrow P$ if it applies for all $q \in Q$ to $(T_qi)(T_qQ)$ as a subspace of $T_{i(q)}P$. For example, an immersion $i: Q \longrightarrow P$ is lagrangian iff $i^*\Omega = 0$ and $\dim Q = \frac{1}{2}\dim P$.

If φ is a 1-form on the manifold X, we may consider it as a mapping from X to T^*X; the equation $\varphi^*\omega_X = \varphi$ is then verified. $\varphi(X)$ is a lagrangian submanifold iff $0 = \varphi^*\Omega_X = \varphi^*(-d\omega_X) = -d(\varphi^*\omega_X) = -d\varphi$, i.e. iff φ is a *closed* form. It follows that the lagrangian submanifolds of T^*X which project diffeomorphically onto X are in natural 1-1 correspondence with the closed 1-forms on X. If the 1-form corresponding to a lagrangian submanifold is of the form dS, S is called a *generating function* for the submanifold. In some sense, we may think of arbitrary lagrangian submanifolds as being "generalized functions" on X. (We will take up this point of view again in Lectures 6 and 10.)

Lagrangian submanifolds also arise in a natural way from symplectomorphisms. We first describe some constructions on symplectic manifolds. If (P, Ω) is any symplectic manifold, $(P, -\Omega)$ is another symplectic manifold called its *dual*. If (P_1, Ω_1) and (P_2, Ω_2) are symplectic manifolds, their ("tensor") *product* $(P_1, \Omega_1) \times (P_2, \Omega_2)$ is defined to be $(P_1 \times P_2, \pi_1^*\Omega_1 + \pi_2^*\Omega_2)$, where $\pi_i: P_1 \times P_2 \longrightarrow P_i$ is the cartesian projection. Now let $f: P_2 \longrightarrow P_1$ be a diffeomorphism, $\Gamma_f = \{(f(x), x) \mid x \in P_2\} \subseteq P_1 \times P_2$ its graph. It is easy to check that Γ_f is a lagrangian submanifold of $(P_1, \Omega_1) \times (P_2, -\Omega_2)$ if and only if f is a symplectomorphism. Arbitrary lagrangian submanifolds of $(P_1, \Omega_1) \times (P_2, -\Omega_2)$ are called *canonical relations*; they may exist even if the dimensions of P_1 and P_2 are unequal. For example, if $f: X \longrightarrow Y$ is a differentiable mapping, the set

$$T^*f = \{((x, \xi), (y, \eta)) \in T^*X \times T^*Y \mid y = f(x), T_x^*\eta = \xi\}$$

is a canonical relation from T^*Y to T^*X. It is the graph of a symplectomorphism if and only if f is a diffeomorphism.

If we happen to have a symplectomorphism between $(P_1, \Omega_1) \times (P_2, -\Omega_2)$ and (T^*X, Ω_X) for some manifold X, we can identify certain canonical relations from P_2 to P_1 with functions on X. This is precisely what is done in the classical theory of generating functions for canonical transformations of $\mathbf{R}^{2n}$ (see **[GO]**); we will see a globalization of this idea in Lecture 7.

It is an interesting problem in differentiable topology to determine which manifolds admit lagrangian immersions and embeddings into a given symplectic manifold. Necessary conditions for there to exist a lagrangian immersion $i: Q \longrightarrow P$ homotopic to a given mapping $j: Q \longrightarrow P$ are:

(i) the de Rham cohomology class $j^*[\Omega] \in H^2(Q; \mathbf{R})$ is zero;

(ii) there is an injective bundle mapping from TQ onto a lagrangian subbundle of the pulled back bundle $j^*(TP)$.

Gromov **[GRM]** and Lees **[LS]** have shown that these two conditions, which are purely homotopy-theoretic, are also sufficient, thus extending the Smale-Hirsch theorem (for plain immersions) **[H]**. We shall see in the next lecture that condition (ii) is equivalent to (ii′) $j^*(TP)$ is isomorphic as a symplectic vector bundle to $TQ \oplus T^*Q$. In particular, Q admits a lagrangian immersion into $\mathbf{R}^{2n}$ if and only if $TQ \oplus T^*Q$ is trivial as a symplectic vector bundle, or, equivalently, if $TQ \otimes_{\mathbf{R}} \mathbf{C}$ is trivial as a complex vector bundle. A sufficient, but not necessary, condition for this is that $TQ \oplus \mathbf{R}$ be trivial, e.g. if Q is a hypersurface in $\mathbf{R}^{n+1}$. (In Lecture 6, we will exhibit explicit immersions for spheres.)

The problem of removing the self-intersections of a lagrangian immersion to obtain an embedding has many ramifications – the question of when this can be done is not yet settled. A special case, discussed in Lecture 7, is the problem of removing the intersection points between a pair of embedded lagrangian submanifolds.

Lecture 2. Lagrangian Splittings, Real and Complex Polarizations, Kähler Manifolds

Every symplectic vector space V contains a lagrangian subspace: Simply choose a maximal element of the lattice of isotropic subspaces of V; if $W \subseteq V$ is isotropic but not lagrangian, we can enlarge it by adjoining an element of $W^{\perp} \setminus W$. Furthermore, if $L \subseteq V$ is a fixed lagrangian subspace, a maximal element in the lattice of isotropic subspaces W such that $L \cap W = \{0\}$ is a lagrangian complement for L. In fact, if $L \cap W = \{0\}$, then $L + W^{\perp} = V$. If $L + W$ is already equal to V, then $L \oplus W = V$, and W is lagrangian. Otherwise, there is an element of $W^{\perp}$ which does not belong to $L + W$; adjoining this element to W moves us higher up in the lattice.

If $L \subseteq V$ is any lagrangian subspace, there is a natural mapping $\alpha : L \longrightarrow (V/L)^*$ given by $\alpha(x)(y + L) = \Omega(x, y)$ for $x \in L$, $y + L \in V/L$. The kernel of α is $\{0\}$; since $\dim L = \dim (V/L)^*$ $(= \frac{1}{2}\dim V)$, α is an isomorphism.

If we have a splitting $V = H \oplus L$ into lagrangian subspaces, the map α just described induces an isomorphism $\beta : L \longrightarrow H^*$; one may now check that the isomorphism $1 \oplus \beta$: $H \oplus L \longrightarrow H \oplus H^*$ is a symplectomorphism from (V, Ω) to $(H \oplus H^*, \Omega_H)$. If $e_1, \ldots, e_n$ is a basis of H, $g_1, \ldots, g_n$ the corresponding dual basis in H^*, and $f_i = \beta^{-1}(g_i)$, we have

$$\Omega(e_i, e_j) = \Omega(f_i, f_j) = 0 \qquad \Omega(e_i, f_j) = -\,\Omega(f_j, e_i) = \delta_{ij}. \tag{2.4.1}$$

A basis of V satisfying equations (2.4.1) is called a *canonical basis* for (V, Ω); it is the counterpart in symplectic geometry of the orthogonal basis in riemannian geometry.

A symplectic vector bundle $E \xrightarrow{\pi} X$ does not, in general, contain a lagrangian subbundle L. For instance, the tangent bundle of an orientable surface other than the torus does not have any 1-dimensional subbundles at all. (We shall see a less trivial example shortly.) It is useful, therefore, to have a generalization which we shall now describe.

Everything we have said up to this point about symplectic vector spaces and bundles remains true if the real field is replaced by $\mathbf{C}$ (or any other field). In particular, if (V, Ω) is a real symplectic vector space, we can extend Ω to a complex-bilinear form $\Omega_{\mathbf{C}}$ on the complexification $V_{\mathbf{C}}$; $(V_{\mathbf{C}}, \Omega_{\mathbf{C}})$ is then a complex symplectic vector space. If $L \subseteq V$ is a lagrangian subspace, $L_{\mathbf{C}}$ is a lagrangian subspace of $V_{\mathbf{C}}$; in general, a lagrangian subspace of $V_{\mathbf{C}}$, which may or may not be of the form $L_{\mathbf{C}}$, is called a *polarization* of V.

A polarization $M \subseteq V_{\mathbf{C}}$ is of the form $L_{\mathbf{C}}$ if and only if $M = \overline{M}$. (If we think of $V_{\mathbf{C}}$ as $V \oplus iV$, then, for $x, y \in V$, $\overline{x + iy}$ is $x - iy$.) At the other extreme are the polarizations N such that $N \cap \overline{N} = \{0\}$, which are called *totally complex*. (We shall discuss the intermediate case in the next lecture.)

If N is totally complex, then $N \cap iV = N \cap V = \{0\}$, so N is the graph of an isomorphism from V to iV, i.e. $N = \{x + iJx \mid x \in V\}$, where $J: V \longrightarrow V$. Since N is a *complex linear* subspace, we have $i(x + iJx) = -Jx + ix \in N$ for all $x \in V$, i.e. $x = J(-Jx)$, or $J^2x = -x$; J must be a *complex structure* on V. Since N is lagrangian, we have for $x, y \in V$

$$\begin{aligned} 0 &= \Omega_{\mathbf{C}}(x + iJx, y + iJy) \\ &= \Omega(x, y) - \Omega(Jx, Jy) + i[\Omega(x, Jy) - \Omega(y, Jx)]. \end{aligned} \tag{2.4.2}$$

The real part of equation (2.4.2) states that J is a symplectomorphism.

Conversely, if J is any symplectomorphism from V to itself such that $J^2 = -$ Identity, reversing the calculations above shows that $\{x + iJx \mid x \in V\} \subseteq V_{\mathbf{C}}$ is a totally complex polarization of V.

We can further investigate a totally complex polarization with the aid of the bilinear form $G(x, y) = \Omega(x, Jy)$, which, according to the imaginary part of equation (2.4.2), is symmetric. If $G(x, y) = 0$ for all y, then $\Omega(x, y) = 0$ for all y and $x = 0$, so G is nondegenerate. Also, we have

$$G(Jx, Jy) = \Omega(Jx, J^2y) = -\Omega(Jx, y) = \Omega(y, Jx) = G(y, x) = G(x, y).$$

Finally, we can recover Ω from J and G by the formula

$$\Omega(x, y) = \Omega(y, J^2x) = G(y, Jx) = G(Jx, y).$$

Any pair (J, G) where J is a complex structure on V and G is a J-invariant nondegenerate symmetric bilinear form is called a *pseudo-hermitian* structure on V (*hermitian* if G is positive definite). The form $\Omega(x, y) = G(Jx, y)$ is called its *associated symplectic structure.* Starting with this Ω and J, we recover G by the rule $G(x, y) = \Omega(x, Jy)$.

To summarize this circle of calculations, we may state that there is a 1-1 correspondence between pseudo-hermitian structures on V and pairs (Ω, N), where Ω is a symplectic structure and N is a totally complex polarization. The polarization N is called *positive* if the associated G is positive definite.

Passing now to vector bundles we will show, following **[ST]**, that every symplectic bundle (over a paracompact base) admits a positive polarization. If (E, Ω) is any symplectic vector bundle, we choose any inner product $\langle\ ,\ \rangle$ on E; Ω is then represented by a skew-adjoint operator $K: E \longrightarrow E$; i.e. $\Omega(x, y) = \langle Kx, y\rangle$. If $K^2 = -$ Identity, we are done, otherwise we consider the polar decomposition $K = RJ$, where $R = \sqrt{KK^t}$ is positive definite symmetric, $J = R^{-1}K$ is orthogonal, and $JR = RJ$. From $K^t = -K$, we can deduce that $J^t = -J$, so $J^2 = -JJ^t = -$ Identity. Also, $\Omega(Jx, Jy) = \langle KJx, Jy\rangle = \langle JKx, Jy\rangle = \langle Kx, y\rangle = \Omega(x, y)$, so J is a symplectomorphism and, therefore, defines a totally complex polarization. Finally, $G(x, y) = \Omega(x, Jy) = \langle Kx, Jy\rangle = \langle JRx, Jy\rangle = \langle Rx, y\rangle$; since R is positive definite, the polarization is positive.

As one application of the existence of positive polarizations, we may conclude that every symplectic vector bundle admits a complex structure; every symplectic manifold can be made into an almost complex manifold. The homotopy class of this structure is determined by the symplectic structure. In fact, if N_0 and N_1 are totally positive polarizations, we may connect the quadratic forms G_0 and G_1 by a straight line $\{G_t\}$. Applying the polar decomposition process above to Ω and the G_t's, we obtain a family $\{J_t\}$ of almost complex structures joining J_0 to J_1.

Another application of positive polarizations concerns lagrangian subbundles: If $L \subseteq E$ is a lagrangian subbundle, then JL is the G-orthogonal complement of L and is again lagrangian. It follows that, if $L \subseteq E$ is a lagrangian subbundle, then it has a lagrangian complement; furthermore, E considered as a complex vector bundle is isomorphic to the complexification of L. (As a symplectic vector bundle, E is isomorphic to $L \oplus L^*$ with the canonical structure Ω_L.)

The arguments above work in the opposite direction as well. Every hermitian vector bundle E has a naturally associated symplectic structure which admits a lagrangian subbundle if and only if E is the complexification $L_{\mathbf{C}}$ of a real bundle. If $E = L_{\mathbf{C}}$, all the odd real Chern classes of E are zero [HI]. Using this fact, we can now find a symplectic vector bundle E which admits a subbundle of dimension ½dim E but no lagrangian subbundle. Let $E_1 \longrightarrow S^2$ be the tangent bundle of S^2, with its symplectic structure given by the area element; the first Chern class of E_1 is twice the generator of $H^2(S^2, \mathbf{R})$. Let $E_2 \longrightarrow S^2$ be the trivial 2-dimensional bundle $S^2 \times \mathbf{R}^2$, and set $E = E_1 \oplus E_2$. Then E is trivial as a real vector bundle, but its first Chern class is equal to that of E_1, hence nonzero, so E admits no lagrangian subbundle.

We look now at polarizations of symplectic manifolds. If (P, Ω) is a symplectic manifold, a lagrangian subbundle L of TP is called a (real) polarization of P if it is involutive; i.e. if the space of sections of L is closed under the Lie bracket operation. In this case, L is the tangent bundle along the leaves of a foliation L of P whose leaves are all lagrangian submanifolds; i.e. a real polarization of P is just a "lagrangian foliation".

The fundamental example of a real polarization is the foliation of a cotangent bundle by its fibres. It is known that every lagrangian foliation is locally symplectomorphic to the foliation of $\mathbf{R}^{2n}$ by the manifolds x_i = constant, and that the leaves of a lagrangian foliation carry a natural flat torsion-free affine connection. (See [W 2].)

A totally complex polarization of TP is called a (totally complex) polarization of P if, as in the real case, it is involutive. Here, this means that the almost complex structure J is integrable – it is a complex structure. Conversely, if we are given a complex manifold (P, J) together with a pseudo-hermitian metric G, we get an associated symplectric structure Ω on TP. A straightforward calculation shows that Ω is a symplectic structure on P, i.e. $d\Omega = 0$, if and only if J is parallel with respect to the Levi-Civita connection of G, i.e. if and only if (P, J, G) is a pseudo-Kähler manifold. This fact yields for us a large new class of symplectic manifolds, including all the nonsingular projective and affine complex algebraic varieties.

For some time, it was suspected that every compact symplectic manifold might have an underlying Kähler structure, or at least that a symplectic manifold might have to satisfy the Hodge relations on its Betti numbers (an incorrect "proof" of this last assertion was published in the early 1950's). Finally, in 1971, W. Thurston produced an example of a four-dimensional symplectic manifold (P, Ω) with $b_1(P) = 3$; since b_{odd} of any Kähler manifold is even, P admits no Kähler structure.

Thurston's example is a flat bundle whose base and fibre are 2-tori. It is simplest to describe the manifold as a quotient of $\mathbf{R}^4$, with coordinates (q_1, p_1, q_2, p_2) and symplectic structure $dq_1 \wedge dp_1 + dq_2 \wedge dp_2$, by the discrete group Γ generated by the symplectomorphisms:

$$\begin{aligned}
a&: (q_1, p_1, q_2, p_2) \mapsto (q_1, p_1, q_2 + 1, p_2),\\
b&: (q_1, p_1, q_2, p_2) \mapsto (q_1, p_1, q_2, p_2 + 1),\\
c&: (q_1, p_1, q_2, p_2) \mapsto (q_1 + 1, p_1, q_2, p_2),\\
d&: (q_1, p_1, q_2, p_2) \mapsto (q_1, p + 1, q_2 + p_2, p_2).
\end{aligned}$$

The "twist" is provided by the map d. The abelianized group $\Gamma/[\Gamma, \Gamma]$ has rank 3. Since $\pi_1(\mathbf{R}^4/\Gamma) = \Gamma$, we have $H^1(\mathbf{R}^4/\Gamma; \mathbf{Z}) = \Gamma/[\Gamma, \Gamma]$, and $b_1(\mathbf{R}^4/\Gamma) = 3$.

After this lecture was presented at the conference, J. Brezin remarked that $\mathbf{R}^4/\Gamma$ is a nilmanifold. If we give $\mathbf{R}^4$ the nilpotent group structure obtained by identifying (q_1, p_1, q_2, p_2) with the matrix

$$\begin{pmatrix} 1 & p_1 & q_2 & 0 & 0 \\ 0 & 1 & p_2 & 0 & 0 \\ 0 & 0 & 1 & 0 & 0 \\ 0 & 0 & 0 & 1 & q_1 \\ 0 & 0 & 0 & 0 & 1 \end{pmatrix},$$

then the group Γ consists of the left translations by the subgroup $\mathbf{Z}^4$. The group $\mathbf{R}^4$ acts transitively on $\mathbf{R}^4/\Gamma$ by right translations, but some of these translations do not preserve the symplectic structure.

Lecture 3. Reduction, the Calculus of Canonical Relations, Intermediate Polarizations

We begin by defining the reduction operation in our three categories.

In (A), if W is any subspace of the symplectic vector space (V, Ω), the quotient $W/W \cap W^\perp$ inherits a symplectic structure in a natural way (see, for example [AT]). When W is coisotropic, the "reduced" symplectic space is just $W/W^\perp$; we denote it by V_W.

If $L \subseteq V$ is lagrangian, it is obvious that the image $L_W = L \cap W/L \cap W^\perp$ of L in V_W is isotropic; in fact it is actually lagrangian as well, regardless of the dimension of $L \cap W$. (To see this we show that, if $x \in W$ is orthogonal to $L \cap W$, then x lies in $(L \cap W) + W^\perp$. Since $(L \cap W)^\perp = L^\perp + W^\perp = L + W^\perp$, we can write $x = l + w^\perp$, where $l \in L$ and $w^\perp \in W^\perp$. But $W^\perp \subseteq W$, so $l = x - w^\perp \in W$; thus $l \in L \cap W$, and $x \in (L \cap W) + W^\perp$.) The operation $L \longmapsto L_W$, which maps the set $L(V)$ of lagrangian subspaces of V to the set $L(V_W)$ is called *reduction relative to W*. It is surjective: any lagrangian subspace in V_W is the reduction of its inverse image in W, which is lagrangian as a subspace of V.

In (A/G), if F is a subbundle of the symplectic vector bundle (E, Ω) such that $F \cap F^\perp$ is a bundle (i.e. the fibres all have the same dimension), then $F/F \cap F^\perp$ becomes a symplectic vector bundle in a natural way. In particular, if F is isotropic, then $F \cap F^\perp = F^\perp$ is certainly a bundle, and the reduced bundle $E_F = F/F^\perp$ exists.

If F is coisotropic and L is a lagrangian subbundle such that $L \cap F$ is a bundle, then $L \cap F^\perp$ is also a bundle (because $L \cap F^\perp = (L + F)^\perp$, and $L + F$ is a bundle whenever $L \cap F$ is), and the reduced bundle $L_F = L \cap F/L \cap F^\perp \subseteq E_F$ exists and is a lagrangian subbundle.

In (G), let Q be a submanifold of the symplectic manifold (P, Ω), and suppose that $TQ \cap TQ^\perp$ is a subbundle of T_QP. Then $TQ/TQ \cap TQ^\perp$ is a symplectic vector bundle; this suggests that Q might project onto a reduced symplectic *manifold.* In fact this is true, at least locally. Our first step in this direction is to show that $TQ \cap TQ^\perp$, considered as a subbundle of TQ, is *involutive.* To see this, we consider vector fields ξ_1 and ξ_2 on Q which are sections of $TQ^\perp$, and we will show that $[\xi_1, \xi_2]$ is a section of $TQ^\perp$. Denote by Ω_Q the pullback of Ω to Q; it is a closed form, and $\Omega_Q(\xi_i, \zeta) = 0$ for any vector field ζ on Q. Now we have, for any vector field η on Q,

$$0 = d\Omega_Q(\xi_1, \xi_2, \eta) = \xi_1\Omega_Q(\xi_2, \eta) - \xi_2\Omega_Q(\xi_1, \eta) + \eta\Omega_Q(\xi_1, \xi_2)$$
$$- \Omega_Q([\xi_1, \xi_2], \eta) + \Omega_Q([\xi_1, \eta], \xi_2) - \Omega_Q([\xi_2, \eta], \xi_1).$$

The first, second, third, fifth, and sixth terms in the sum above are zero since they are of the form $\Omega_Q(\xi_i, \zeta)$, so the fourth term must be zero as well. Since η was arbitrary, it follows that $[\xi_1, \xi_2]$ is a section of $TQ^\perp$.

We see that $TQ \cap TQ^{\perp}$ is the tangent bundle to a *foliation* of Q which we denote by $\underset{\sim}{Q}^{\perp}$. Locally (and globally, if the foliation $\underset{\sim}{Q}^{\perp}$ is sufficiently nice), we may form the quotient manifold $Q/\underset{\sim}{Q}^{\perp}$. We note that the tangent space to $Q/\underset{\sim}{Q}^{\perp}$ at one of its points (i.e. at a leaf of $\underset{\sim}{Q}^{\perp}$) may be identified with the fibre of the quotient bundle $TQ/TQ \cap TQ^{\perp}$ at any point of this leaf. This gives us a symplectic structure on the tangent space of $Q/\underset{\sim}{Q}^{\perp}$; to show that this structure is independent of the point chosen on the leaf, we observe that the Lie derivative $L_{\xi}\Omega_Q$ is zero for any section ξ of $TQ \cap TQ^{\perp}$. In fact, $L_{\xi}\Omega_Q = d(\xi \lrcorner \Omega_Q) + \xi \lrcorner d\Omega_Q$; the first term is zero because ξ lies in $TQ^{\perp}$, and the second is zero because $d\Omega_Q = 0$. It follows that there is a naturally determined symplectic structure on $Q/\underset{\sim}{Q}^{\perp}$ whose pullback to Q is the form Ω_Q. (If P_Q does not exist as a manifold, the proper interpretation of the fact $L_{\xi}\Omega_Q = 0$ is that the *holonomy* of the foliation $\underset{\sim}{Q}^{\perp}$ is contained in the pseudo-group of symplectomorphisms **[LN]**.)

If Q is a coisotropic submanifold of P, then $\underset{\sim}{Q}^{\perp}$ always exists: it is a foliation of Q by isotropic submanifolds. The reduced manifold $Q/\underset{\sim}{Q}^{\perp}$ is denoted by P_Q; it is defined at least locally. The subset $\{(x, y) \mid y \in Q, y \text{ belongs to the leaf } x\} \subseteq P_Q \times P$ is called the graph of the reduction by Q; it is a canonical relation from P to P_Q. (Proving this statement amounts to proving it in the category (A). See **[SN]**.)

We now wish to consider the possibility of reducing a lagrangian submanifold L of P to obtain a lagrangian submanifold L_Q of P_Q. The first step is to look at the intersection $L \cap Q$ and the subset $TL \cap TQ$ of $T_{L \cap Q}P$. We say that L and Q have *clean intersection* (this concept goes back to **[BO]**) if $L \cap Q$ is a manifold and $TL \cap TQ = T(L \cap Q)$. In this case, $TL \cap TQ$ is a bundle, and $TL \cap TQ^{\perp}$ is a bundle as well. Since $TL \cap TQ^{\perp}$ is exactly the kernel of the differential of the restriction to $L \cap Q$ of the projection $Q \longrightarrow Q/\underset{\sim}{Q}^{\perp} = P_Q$, it follows that the projection of $L \cap Q$ into P_Q has constant rank; the image is therefore an "immersed submanifold" $L_Q \subseteq P_Q$ which must, by our vector bundle considerations, be lagrangian. A special case of clean intersection is that of transverse intersection, i.e. when $T_{L\cap Q}L + T_{L \cap Q}Q = T_{L \cap Q}P$. In this case, $TL \cap TQ^{\perp}$ consists only of zero vectors, and the manifold $L \cap Q$ immerses into P_Q. (The immersion may not be an embedding because $L \cap Q$ may intersect a leaf of $\underset{\sim}{Q}^{\perp}$ in several points.)

An interesting example of the reduction construction is provided by the unit sphere $S = S^{2n-1} = \{(x, \xi) \mid \Sigma_{i=1}^n (x_i^2 + \xi_i^2) = 1\}$, which is a coisotropic submanifold of $\mathbf{R}^{2n}$. The tangent space $T_{(x,\xi)}S$ consists of those vectors $\Sigma_{i=1}^n (a_i \partial/\partial x_i + b_i \partial/\partial \xi_i)$ for which $\Sigma_{i=1}^n (x_i a_i + \xi_i b_i) = 0$. One may check that the orthogonal space is generated by the vector $\Sigma_{i=1}^n (\xi_i \partial/\partial x_i - x_i \partial/\partial \xi_i)$ (we shall see in the next section where this vector comes from), so the leaves of the foliation $\underset{\sim}{S}^{\perp}$ are the solution curves of the differential equations $dx_i/dt = \xi_i$ and $d\xi_i/dt = -x_i$. If we identify $\mathbf{R}^{2n}$ with $\mathbf{C}^n$ by introducing the complex variables $z_i = x_i + \sqrt{-1}\,\xi_i$, these equations become $dz_i/dt = -\sqrt{-1}\, z_i$, and the solution curves are given by

$$z_i(t) = e^{-\sqrt{-1}\,t} z_i(0).$$

Each solution curve is a circle–the intersection of S^{2n-1} with a complex line in $\mathbf{C}^n$. The quotient $S/\underset{\sim}{S}^{\perp}$ may thus be identified with the complex projective space $\mathbf{C}P^{n-1}$ of lines in $\mathbf{C}^{n-1}$; this gives a "symplectic" construction of a symplectic structure on $\mathbf{C}P^{n-1}$.

(It can be shown to be the same as the one which comes from its Kähler metric.)

The x-"axis" given by $\xi_1 = \cdots = \xi_n = 0$ is a lagrangian submanifold L in $\mathbf{R}^{2n}$. $L \cap S^{2n-1}$ is the unit sphere S^{n-1} in $\mathbf{R}^n$; it intersects each leaf of $S^\perp$ in two antipodal points or not at all, so the lagrangian immersion of $L \cap S^{2n-1}$ into CP^{n-1} is a double covering of the image, which is RP^{n-1} embedded as a lagrangian submanifold of CP^{n-1}.

More generally, if $P \subseteq CP^{n-1}$ is a nonsingular algebraic variety defined by polynomials with real coefficients, then the real points of P form a lagrangian submanifold of P (with the symplectic structure induced from CP^{n-1}).

Perhaps the most important application of the reduction operation is to the composition of canonical relations. If (P_i, Ω_i) $(i = 1, 2, 3)$ are symplectic manifolds, and $f: P_2 \longrightarrow P_1$ and $g: P_3 \longrightarrow P_2$ are symplectomorphisms, the composition $f \circ g: P_3 \longrightarrow P_1$ is again a symplectomorphism. Suppose now that $C \subseteq (P_1, \Omega_1) \times (P_2, -\Omega_2)$ and $D \subseteq (P_2, \Omega_2) \times (P_3, -\Omega_3)$ are canonical relations. The composition $C \circ D$ is defined as

$$\{(p_1, p_3) \in P_1 \times P_3 \mid \exists\, p_2 \in P_2 \text{ such that } (p_1, p_2) \in C \text{ and } (p_2, p_3) \in D\}.$$

To find conditions under which $C \circ D$ is again a canonical relation, we observe that it can be obtained by forming the product $C \times D \subseteq P_1 \times P_2 \times P_2 \times P_3$, intersecting with $P_1 \times \Delta_{P_2} \times P_3$, where $\Delta_{P_2} \subseteq P_2 \times P_2$ is the diagonal, and projecting onto $P_1 \times P_3$. It is easy to check that $\Delta = P_1 \times \Delta_{P_2} \times P_3$ is a coisotropic submanifold of $(P, \Omega) = (P_1, \Omega_1) \times (P_2, -\Omega_2) \times (P_2, \Omega_2) \times (P_3, -\Omega_3)$, and that the reduced symplectic manifold P_Δ may be identified with $(P_1, \Omega_1) \times (P_3, -\Omega_3)$. Furthermore, $C \times D$ is a lagrangian submanifold of (P, Ω), and its reduction $(C \times D)_\Delta$ is just $C \circ D$. It follows that, if the intersection of $C \times D$ with Δ is clean, then $C \circ D$ is a (possibly immersed) canonical relation from P_3 to P_1. In particular, this is the case whenever $C \circ D$ is transversal to Δ.

If P_3 is the zero-dimensional manifold consisting of a single point, then a canonical relation from P_3 to P_2 is just a lagrangian submanifold L of P_2. If C is a canonical relation from P_2 to P_1, the composition $C \circ L$ is just the *image* $C(L) \subseteq P_1$. It follows that, under the assumption of clean intersection, we can operate on lagrangian submanifolds with canonical relations. (At this point, we could close a circle of reasoning by observing that the reduction of lagrangian submanifolds is just operation by the graph of the reduction.)

For example, if $f: X \longrightarrow Y$ is a differentiable mapping, then the canonical relation T^*f (see Lecture 1) operates on certain lagrangian submanifolds of T^*Y to give lagrangian submanifolds of T^*X. In particular, if $L \subseteq T^*Y$ has a generating function $S: Y \longrightarrow \mathbf{R}$, then $(T^*f)(L) \subseteq T^*X$ has $S \circ f$ as its generating function; this reinforces our assertion that lagrangian submanifolds of cotangent bundles should be thought of as generalized functions. If f is a submersion, then any lagrangian submanifold of T^*Y may be operated upon by T^*f (the intersection is transversal); if f is an immersion, then the inverse relation $(T^*f)^{-1}$ from T^*X to T^*Y operates on all lagrangian submanifolds of T^*X. This behavior is analogous to that of distributions, which can always be pulled back under submersions and pushed forward under immersions.

We can use the reduction operation to give a geometric interpretation of "intermediate" polarizations; i.e. those which are neither real nor totally complex. In the category (A), if

$M \subseteq V_{\mathbf{C}}$ is any polarization, then $M + \bar{M}$ is the complexification of a coisotropic subspace $M_{\mathbf{R}}$ of V. We leave the verification of the following facts to the reader:

(i) $M \cap \bar{M} = (M_{\mathbf{R}}^{\perp})_{\mathbf{C}}$;

(ii) M determines a totally complex polarization on the reduced space $M_{\mathbf{R}}/M_{\mathbf{R}}^{\perp}$;

(iii) If $W \subseteq V$ is any coisotropic subspace, every totally complex polarization on $W/W^{\perp}$ arises from a unique polarization M on V with $M_{\mathbf{R}} = W$.

In category (A/G), we will reserve the term *polarization* of E for those lagrangian subbundles $B \subseteq E_{\mathbf{C}}$ for which $B \cap \bar{B}$ is a bundle, i.e. has constant fibre dimension. Then there is a natural 1-1 correspondence between polarizations of E and pairs consisting of a coisotropic subbundle $F \subseteq E$ and a totally complex polarization on $F/F^{\perp}$.

In category (G), let (P, Ω) be a symplectic manifold $F \subseteq T_{\mathbf{C}}P$ a polarization in the sense of bundles. Before we can call F a polarization of P, we must assume that it is involutive, but that is not enough. The "real part" $F_{\mathbf{R}}$ of F is a coisotropic subbundle of TP, and we must assume separately that $F_{\mathbf{R}}$ is involutive. If all these conditions are satisfied, we call F a polarization of P, and we have the following structures:

a coistropic foliation $\mathcal{F}$ on P

the isotropic foliation $\mathcal{F}^{\perp} \subseteq \mathcal{F}$

the foliation $\mathcal{F}/\mathcal{F}^{\perp}$ of $P/\mathcal{F}^{\perp}$, each of whose leaves has a symplectic structure

a totally complex polarization (pseudo-Kähler structure) on each leaf of $\mathcal{F}/\mathcal{F}^{\perp}$.

For further discussion and examples, we refer the reader to **[SIM]**.

Lecture 4. Hamiltonian Systems and Group Actions on Symplectic Manifolds

If $f\colon P \longrightarrow \mathbf{R}$ is a C^∞ function and (P, Ω) is symplectic, the *hamiltonian vector field* H_f on P is defined as $\Omega^{-1} \circ df$; i.e. if η is any vector field on P, we have $\Omega(H_f, \eta) = (\widetilde{\Omega} \circ H_f)(\eta) = df(\eta) = \eta \cdot f$. We may derive easily the fundamental properties

$$H_f \cdot f = \Omega(H_f, H_f) = 0$$

and

$$L_{H_f}\Omega = d(H_f \lrcorner \Omega) + H_f \lrcorner d\Omega = d(df) + H_f \lrcorner 0 = 0.$$

In other words, the flow generated by H_f leaves the function f and the symplectic structure Ω invariant. f is called a *hamiltonian function* for the flow.

If $P = \mathbf{R}^{2n}$ with its standard symplectic structure $\Sigma\, dx_i \wedge d\xi_i$, then $\widetilde{\Omega}(\partial/\partial x_i) = d\xi_i$ and $\widetilde{\Omega}(\partial/\partial \xi_i) = -dx_i$, so

$$H_f = \widetilde{\Omega}^{-1}\left(\sum \frac{\partial f}{\partial x_i} dx_i + \frac{\partial f}{\partial \xi_i} d\xi_i\right) = \sum\left(-\frac{\partial f}{\partial x_i}\frac{\partial}{\partial \xi_i} + \frac{\partial f}{\partial \xi_i}\frac{\partial}{\partial x_i}\right),$$

and the integral curves of H_f are solution curves of the equations

$$\frac{dx_i}{dt} = \frac{\partial f}{\partial \xi_i} \qquad \frac{d\xi_i}{dt} = -\frac{\partial f}{\partial x_i},$$

which are known in mechanics as *Hamilton's equations.*

In fact, as Whittaker **[WH]** and Souriau **[SR]** have already pointed out, "hamiltonian" dynamical systems can already be found in the work of Lagrange **[LA 1]**. In 1808, near the end of a lifetime of work in celestial mechanics, Lagrange discovered that the equations which express the perturbation of elliptical planetary motion due to interactions could be put in a simple, general form.

Considering a single planet, Lagrange observed that the possible elliptic motions which the planet would follow under the influence of the sun alone were described by six real parameters, a, b, c, f, g, h. To describe the variation from elliptic motion caused by the influence of other planets, he proposed to determine the derivatives of a, b, c, f, g, h with respect to time. After obtaining rather complicated relations for these derivatives, he stated:

> "ces six équations donneront les six differentielles *da, db,* ..... par l'elimination ordinaire; mais on aurait de cette manière des formules très-compliquées. Heureusement j'ai trouvé une combinaison de ces équations qui conduit à des résultats simples et très-remarquables, et que je vais exposer."

To this end, he introduced the expressions, now called Lagrange brackets, (a, b), (a, c), $\ldots$, (b, c), $\ldots$, (g, h), which are combinations of derivatives with respect to a, b, c, f, g, h of the position and velocity at a fixed time t_0.

He observed that these brackets are independent of the choice of t_0. Next, introducing a certain "disturbing function" Ω, which depends on a, b, c, f, g, h (and, in a slowly varying way, on the time), he wrote the equations

$$\frac{d\Omega}{da}dt = (a, b)\,db + (a, c)\,dc + (a, f)\,df + (a, g)\,dg + (a, h)\,dh$$

$$\frac{d\Omega}{db}dt = -(a, b)\,da + (b, c)\,dc + (b, f)\,df + (b, g)\,dg + (b, h)\,dh$$

$$\vdots$$

$$\frac{d\Omega}{dh}dt = \cdot \; - \cdot - (h, g)\,dg.$$

Finally, he remarked that these linear equations may be solved for da/dt, $\ldots$, dh/dt in terms of $d\Omega/da, \ldots, d\Omega/dh$.

In our terminology, the Lagrange brackets are the coefficients of a symplectic structure expressed in terms of the coordinates (a, b, c, f, g, h), and Ω is the "hamiltonian" function. Lagrange's equations correspond to our $\widetilde{\Omega} \circ H_f = df$ (be careful, our f is Lagrange's Ω), and solving the equations corresponds to finding $\widetilde{\Omega}^{-1} \circ df$.

In his next paper [**LA** 2], Lagrange observed that, for a certain choice of coordinates $(\alpha, \beta, \gamma, \lambda, \mu, \nu)$ (where $\lambda = \partial T/\partial\alpha'$, $\mu = \partial T/\partial\beta'$, $\nu = \partial T/\partial\gamma'$, T being the "total energy", $'$ denoting a time derivative), the equations take the form

$$\frac{d\Omega}{d\alpha}dt = d\lambda \qquad \frac{d\Omega}{d\beta}dt = d\mu \qquad \frac{d\Omega}{d\gamma}dt = d\nu$$

$$\frac{d\Omega}{d\lambda}dt = -d\alpha \qquad \frac{d\Omega}{d\mu}dt = -d\beta \qquad \frac{d\Omega}{d\nu}dt = -d\gamma$$

"qui sont, comme l'on voit, sous la forme la plus simple qu'il soit possible." These are exactly "Hamilton's equations".

Hamilton went on to explore an area which Lagrange merely discovered. In particular, Lagrange seems not to have observed that the motions themselves, and not just their variations, could be described by equations in his simple form. Furthermore, Lagrange's use of these equations was mainly computational; he did not explore their theoretical consequences as did Hamilton and Jacobi.

Returning from our historical digression, we give next some applications of hamiltonian vector fields to symplectic geometry. If Q is a regular level surface of $f: P \longrightarrow \mathbf{R}$ (i.e. df is nowhere vanishing on Q, f is constant on Q, and $\dim Q = \dim P - 1$), then the vector field H_f, which is tangent to Q because $H_f \cdot f = 0$, generates the isotropic subbundle $TQ^\perp$. In fact, if η is any vector field tangent to Q, we have $\Omega(H_f, \eta) = \eta \cdot f$, which is zero along Q, so H_f maps Q into $TQ^\perp$. Since $H_f = \widetilde{\Omega}^{-1} \circ df$ vanishes nowhere on Q, and $TQ^\perp$ is 1-dimensional, H_f must generate $TQ^\perp$. (The vector field

$$\sum_{i=1}^{n} \xi_i \left(\frac{\partial}{\partial x_i} - x_i \frac{\partial}{\partial \xi_i}\right)$$

which occurred in the last section was just the Hamiltonian vector field of $\frac{1}{2}(\Sigma_{i=1}^n(x_i^2 + \xi_i^2) - 1)$, which has the sphere S as a regular level surface.)

More generally, if Q is any coisotropic submanifold, we can effectively construct the distribution $TQ^{\perp}$ by choosing (locally) functions $f_1, \ldots, f_k$ such that Q is defined by the equations $f_1 = \cdots = f_k = 0$, and $df_1, \ldots, df_k$ are linearly independent along Q. If η is any vector field tangent to Q, we have $\Omega(H_{f_i}, \eta) = \eta \cdot f_i = 0$ along Q, as before, so H_{f_i} is a section of $TQ^{\perp}$ along Q. Since dim $TQ^{\perp} = k$, the k linearly independent vector fields $H_{f_1}, \ldots, H_{f_k}$ form a spanning set.

Hamiltonian vector fields are almost characterized by the fact that they generate flows which preserve the symplectic structure. In fact, if ξ is any vector field we have

$$L_\xi \Omega = d(\xi \lrcorner \Omega) + \xi \lrcorner d\Omega = d(\xi \lrcorner \Omega),$$

so $L_\xi \Omega = 0$ if and only if $\xi \lrcorner \Omega$ is a closed form. If $\xi \lrcorner \Omega$ is exact (which is always the case if $\xi \lrcorner \Omega$ is closed and $H^1(P; \mathbf{R}) = 0$), i.e. $\xi \lrcorner \Omega = dS$, then ξ is equal to the hamiltonian vector field H_s; S is determined by ξ up to an additive locally constant function. We may express these observations by an exact sequence

$$0 \longrightarrow H^0(P; \mathbf{R}) \xrightarrow{\alpha} C^\infty(P; \mathbf{R}) \xrightarrow{\beta} X(P, \Omega) \xrightarrow{\gamma} H^1(P; \mathbf{R}) \longrightarrow 0,$$

where $X(P, \Omega)$ is the Lie algebra consisting of those vector fields ξ for which $L_\xi \Omega = 0$.

It happens that $C^\infty(P, \mathbf{R})$ has a Lie algebra structure with respect to which the mapping β is a homomorphism. In fact, for functions f and g, we define their *Poisson bracket* $\{f, g\}$ by the formula

$$\{f, g\} = \Omega(H_{f_2}, H_{f_1}) = H_{f_1} \cdot f_2.$$

To see that $H_{\{f,g\}} = [H_f, H_g]$, we compute $L_{\xi_1}(\xi_2 \lrcorner \Omega)$ in two different ways for ξ_1 and ξ_2 in $X(P, \Omega)$. First of all, since Lie differentiation is a derivation of any naturally defined bilinear operation on tensors, we have

$$L_{\xi_1}(\xi_2 \lrcorner \Omega) = L_{\xi_1}\xi_2 \lrcorner \Omega + \xi_2 \lrcorner L_{\xi_1}\Omega = [\xi_1, \xi_2] \lrcorner \Omega + \xi_2 \lrcorner 0.$$

On the other hand, by the usual identity for the Lie derivative on forms,

$$L_{\xi_1}(\xi_2 \lrcorner \Omega) = \xi_1 \lrcorner d(\xi_2 \lrcorner \Omega) + d(\xi_1 \lrcorner \xi_2 \lrcorner \Omega) = \xi_1 \lrcorner 0 + d[\Omega(\xi_2, \xi_1)].$$

Comparing these two calculations, we obtain

$$[\xi_1, \xi_2] \lrcorner \Omega = d[\Omega(\xi_2, \xi_1)];$$

i.e. $[\xi_1, \xi_2]$ is the hamiltonian vector field of the function $\Omega(\xi_2, \xi_1)$. In particular, if $\xi_1 = H_{f_1}$ and $\xi_2 = H_{f_2}$ then $[H_{f_1}, H_{f_2}] = H_{\{f_1, f_2\}}$. We conclude that the map $C^\infty(P; \mathbf{R}) \xrightarrow{\beta} X(P, \Omega)$ is a homomorphism of Lie algebras, and that the commutator algebra $[X(P, \Omega), X(P, \Omega)]$ is contained in the image of α. Another way of expressing the last fact is to say that, if we give $H^1(P; \mathbf{R})$ the trivial Lie algebra structure, in which all brackets are zero, then γ is a homomorphism of Lie algebras. Similarly, since the Poisson bracket of any two locally constant functions is zero, the map α is a Lie algebra homomorphism if $H^0(P; \mathbf{R})$ is given the trivial structure, so our sequence above is actually an exact sequence of Lie algebras.

The study of a single element of $X(P, \Omega)$ is a special case of the study of a homomorphism $\mathfrak{g} \xrightarrow{\rho} X(P, \Omega)$, where $\mathfrak{g}$ is a finite-dimensional Lie algebra; such a homomorphism is called an *action* of $\mathfrak{g}$ on (P, Ω). It is natural to ask when such an action lifts through β to a homomorphism from $\mathfrak{g}$ to $C^\infty(P, \mathbf{R})$. (We shall see later that this question is important in quantization theory.)

As a first step, we may simply ask when ρ lifts through β as a linear mapping. For this, it is sufficient that $\rho(\mathfrak{g})$ be in the image of $\mathfrak{g}$, i.e. $\gamma \circ \rho = 0$. Since $\mathfrak{g} \xrightarrow{\gamma\circ\rho} H^1(P; \mathbf{R})$ is a Lie algebra homomorphism and $H^1(P; \mathbf{R})$ is abelian, it follows that $\gamma \circ \rho$ annihilates $[\mathfrak{g}, \mathfrak{g}]$, so the obstruction to lifting γ is the induced mapping $\mathfrak{g}/[\mathfrak{g}, \mathfrak{g}] \xrightarrow{\bar{\rho}} H^1(P; \mathbf{R})$. There are many interesting situations where this invariant must vanish:

(i) If $\mathfrak{g}/[\mathfrak{g}, \mathfrak{g}] = 0$ (by the "first Whitehead lemma" **[J]**, this is the case whenever $\mathfrak{g}$ is semi-simple).

(ii) If there is a 1-form ω on P such that $-d\omega = \Omega$ and $L_\xi \omega = 0$ for each $\xi \in \rho(\mathfrak{g})$. ($0 = L_\xi\omega = d(\xi \lrcorner \omega) + \xi \lrcorner d\omega$, so $d(\xi \lrcorner \omega) = \widetilde{\Omega} \circ \xi$, and $\xi \lrcorner \omega$ is a hamiltonian function for ξ; this case occurs when $(P, \Omega) = (T^*X, \Omega_X)$ and ρ is the lift of an action of $\mathfrak{g}$ on X.)

(iii) If $H^1(P; \mathbf{R}) = 0$.

A simple situation where $\bar{\rho}$ does not vanish is given by $P = S^1 \times S^1$, $\Omega = d\theta_1 \wedge d\theta_2$, $\mathfrak{g} = \mathbf{R}^2$, $\rho(x_1, x_2) = x_1 \partial/\partial\theta_1 + x_2 \partial/\partial\theta_2$. In this case, $[\mathfrak{g}, \mathfrak{g}] = 0$, and $\bar{\rho}$ is an isomorphism of $\mathfrak{g}/[\mathfrak{g}, \mathfrak{g}]$ onto $H^1(S^1 \times S^1; \mathbf{R})$.

Given a linear map $\sigma : \mathfrak{g} \longrightarrow C^\infty(P, \mathbf{R})$ such that $\beta \circ \sigma = \rho$, it is useful to consider as well the mapping $\mu : P \longrightarrow \mathfrak{g}^*$ defined by $[\mu(p)](v) = [\sigma(v)](p)$ for $p \in P$, $v \in \mathfrak{g}$. μ is called a *momentum function* for the action ρ; it is determined by ρ up to addition by a locally constant function from P to $\mathfrak{g}^*$.

If the action is *transitive* in the sense that the vector fields in $\rho(\mathfrak{g})$ span the tangent bundle of P, then one may check that any momentum function for ρ is an immersion of P into $\mathfrak{g}^*$. This observation leads to an important result of Kirillov **[KI 2]**, Kostant **[KO]**, and Souriau **[SR]** which describes all symplectic homogeneous spaces of a finite dimensional Lie group G in terms of the *coadjoint* action of G on $\mathfrak{g}^*$.

Our "momentum" is so designated because it generalizes the linear and angular momenta of classical mechanics. For instance, if we let $\mathbf{R}^3$ (coordinates (a_1, a_2, a_3)) act on $\mathbf{R}^6$ (coordinates $(x_1, \ldots, \xi_3)$) by $\rho(a_1, a_2, a_3) = a_1\partial/\partial x_1 + a_2 \partial/\partial x_2 + a_3 \partial/\partial x_3$, then a momentum function is given by

$$[\mu(x_1, x_2, x_3, \xi_1, \xi_2, \xi_3)] (a_1, a_2, a_3) = \xi_1 a_1 + \xi_2 a_2 + \xi_3 a_3.$$

In terms of the coordinates on $\mathbf{R}^3$ dual to (a_1, a_2, a_3), we have $\mu(x, \xi) = \xi$; classically, (ξ_1, ξ_2, ξ_3) are the momentum variables conjugate to the coordinates (x_1, x_2, x_3). Next, let $\mathfrak{g} = \mathbf{R}^3$ with the structure given by the cross product. $\mathfrak{g}$ acts on $\mathbf{R}^3 \times \mathbf{R}^3$ by $\rho(v)(x, \xi) = (v \times x, v \times \xi)$. (This is the infinitesimal version of the action of the rotation group $SO(3)$.)

Some calculation shows that the momentum for this action is given by $\mu(x, \xi) = x \times \xi$ (we identify $\mathfrak{g}$ with $\mathfrak{g}^*$ by using the Euclidean inner product on $\mathbf{R}^3$), which is the usual *angular momentum.* For further examples and discussion, see **[SM]** and **[SR]**.

Once we have a lift $\sigma: \mathfrak{g} \longrightarrow C^\infty(P, \mathbf{R})$ of the action $\rho: \mathfrak{g} \longrightarrow X(P, \mathbf{R})$, we may go on to ask whether it is a Lie algebra homomorphism. Given elements v_1 and v_2 in $\mathfrak{g}$, the element $\widetilde{\sigma}(v_1, v_2) = \sigma[v_1, v_2] - \{\sigma(v_1), \sigma(v_2)\}$ of $C^\infty(P, \mathbf{R})$ goes to zero under β because $\rho = \beta \circ \sigma$ is a homomorphism, so it must be locally constant. Thus, $\widetilde{\sigma}$ is a map from $\mathfrak{g} \times \mathfrak{g}$ to $H^0(P; \mathbf{R})$. $\widetilde{\sigma}$ is obviously skew-symmetric; the additional identity $\widetilde{\sigma}([x, y], z) + \widetilde{\sigma}([y, z], z) + \widetilde{\sigma}([z, x], y) = 0$ follows from the Jacobi identities in and $C^\infty(P, \mathbf{R})$. Such a bilinear map is called a *2-cocycle* on $\mathfrak{g}$ with coefficients in $H^0(P; \mathbf{R})$–it measures the degree to which σ fails to be a homomorphism.

Recall, now, that σ was determined by ρ only up to addition of a map $\theta: \mathfrak{g} \longrightarrow H^0(P, \mathbf{R})$. If we replace σ by $\sigma + \theta$, we obtain

$$\begin{aligned}
\widetilde{(\sigma + \theta)}(v_1, v_2) &= (\sigma + \theta)[v_1, v_2] - \{(\sigma + \theta)(v_1), (\sigma + \theta)(v_2)\} \\
&= \sigma[v_1, v_2] + \theta[v_1, v_2] - \{\sigma(v_1) + \theta(v_1), \sigma(v_2) + \theta(v_2)\} \\
&= \sigma[v_1, v_2] + \theta[v_1, v_2] - \{\sigma(v_1), \sigma(v_2)\}
\end{aligned}$$

(since $\theta(v_1)$ and $\theta(v_2)$ are in the center of the Lie algebra $C^\infty(P, \mathbf{R})$). Thus, $\widetilde{(\sigma + \theta)}(v_1, v_2) = \widetilde{\sigma}(v_1, v_2) + \theta[v_1, v_2]$.

The map $(v_1, v_2) \longmapsto \theta[v_1, v_2]$ is a 2-cocycle of a special type, called a 2-coboundary. The map $\widetilde{\sigma}$, considered as an element of 2-cocycles/2-coboundaries, is independent of the choice of σ, so we denote it by $\widetilde{\rho}$; it is the obstruction to finding a lift of ρ to a Lie algebra homomorphism.

Once again, there are some cases in which one can predict that $\widetilde{\rho}$ is zero:

(i) If every 2-cycle on $\mathfrak{g}$ is a 2-coboundary (by the "second Whitehead lemma" **[J]**, this is the case whenever $\mathfrak{g}$ is semi-simple);

(ii) If there is a 1-form ω on P such that $-d\omega = \Omega$ and $L_\xi \omega = 0$ for each $\xi \in \rho(\mathfrak{g})$. (If we put $\sigma(v) = \rho(v) \lrcorner \omega$, we have

$$\begin{aligned}
\sigma[v_1, v_2] &= [\rho(v_1), \rho(v_2)] \lrcorner \omega = L_{\rho(v_1)}(\rho(v_2) \lrcorner \omega) \quad (\text{since } L_{\rho(v_1)}\omega = 0) \\
&= \rho(v_1) \lrcorner d(\rho(v_2) \lrcorner \omega) = \rho(v_1) \lrcorner \{L_{\rho(v_2)}\omega - \rho(v_2) d\omega\} \\
&= \rho(v_1) \lrcorner (\rho(v_2) \lrcorner \Omega) = \Omega(\rho(v_2), \rho(v_1)) \\
&= \Omega(\xi_{\sigma(v_2)}, \xi_{\sigma(v_1)}) = \{\sigma(v_1), \sigma(v_2)\}.)
\end{aligned}$$

The simplest situation in which $\widetilde{\rho}$ does not vanish is given by $\mathfrak{g} = \mathbf{R}^2$ (coordinates (a, b)), $P = \mathbf{R}^2$ (coordinates (x, ξ)), and $\rho(a, b) = a\partial/\partial x + b\partial/\partial\xi$. A lift is given by $\sigma(a, b) = a\xi - bx$, and we have

$$\begin{aligned}
\widetilde{\sigma}((a_1, b_1), (a_2, b_2)) &= \sigma([a_1, b_1], [a_2, b_2]) - \{\sigma(a_1, b_1), \sigma(a_2, b_2)\} \\
&= -\{a_1\xi - b_1 x, a_2\xi - b_2 x\} = a_2 b_1 - a_1 b_2.
\end{aligned}$$

Since $[\mathfrak{g}, \mathfrak{g}] = 0$, the only coboundary is zero, so the obstruction $\widetilde{\rho}$ is not zero.

For actions of the galilean group, the cohomology class represents the *mass* of a physical system; see [SR].

We can interpret some of the results above and obtain further information by looking at symplectic actions of Lie *groups*. A smooth symplectic action $G \times P \xrightarrow{A} P$ of the Lie group G on (P, Ω) gives rise to an action $\mathfrak{g} \xrightarrow{\rho} \mathfrak{X}(P, \Omega)$ of the Lie algebra $\mathfrak{g}$ of G; conversely, every ρ arises *locally* from an action of G, the two possible obstructions to lifting globally to G being the incompleteness of some vector fields in $\rho(\mathfrak{g})$ and an effect of the non-simple-connectivity of G.

Given an action $G \times P \xrightarrow{A} P$, we may consider the closed two-form $\Omega_A = A^*\Omega - \pi_P^*\Omega$ on $G \times P$, where $G \times P \xrightarrow{\pi_P} P$ is the cartesian projection. We consider $G \times P$ itself as a G-space by letting G act on itself by left translations and on P trivially; then the map A is equivariant, i.e. $A(gh, p) = A(g, A(h, p))$, by the definition of a group action. It follows that $A^*\Omega$ is a G-invariant form on $G \times P$. Obviously, $\pi_P^*\Omega$ is G-invariant, so Ω_A is G-invariant.

Now we may consider the class $[\Omega_A]$ in the equivariant cohomology group $H_G^2(G \times P; \mathbf{R})$ of closed modulo exact G-invariant forms. By the Künneth formula (for which I can unfortunately find no reference in this equivariant case, but which I am assured is true), $H_G^2(G \times P; \mathbf{R})$ is naturally isomorphic to

$$[H_G^0(G; \mathbf{R}) \otimes H^2(P; \mathbf{R})] \oplus [H_G^1(G; \mathbf{R}) \otimes H^0(P; \mathbf{R})] \oplus [H_G^2(G; \mathbf{R}) \otimes H^0(P; \mathbf{R})],$$

so we may consider separately the three components of $[\Omega_A]$. (On the level of forms, this amounts to decomposing Ω_A into components of different "bidegree" with respect to the product structure on $G \times P$.) The first component is zero because the restriction of Ω_A to any manifold of the form $\{g\} \times P$ vanishes since A is a symplectic action. (If A were not symplectic, this component of $[\Omega_A]$ would measure the extent to which the action of each component of G failed to preserve $[\Omega]$.) As for the second component, one can identify $H_G^1(G; \mathbf{R})$ with the dual space of $\mathfrak{g}/[\mathfrak{g}, \mathfrak{g}]$; thus, the second component of $[\Omega_A]$ may be interpreted as a map $\mathfrak{g}/[\mathfrak{g}, \mathfrak{g}] \longrightarrow H^1(P; \mathbf{R})$. Direct computation shows that this map is exactly the obstruction $\bar{\rho}$ to the existence of a momentum function, which we discussed above. The vanishing of $\bar{\rho}$, in the present terms, means that we can find a G-invariant 1-form φ on $G \times P$ such that $d\varphi - \Omega_A$ consists only of terms of bidegree (2,0); i.e. $d\varphi(x, y) = 0$ if either x or y is in the "P direction". The G-component of φ along the manifold $\{e\} \times P$ may be considered as a map from P to $T_e^*G = \mathfrak{g}^*$; one may check that this map is a momentum function for ρ. (e is the identity element of G.)

The summand $H_G^2(G; \mathbf{R}) \otimes H^0(P; \mathbf{R})$ consists of closed modulo exact G-invariant forms on G with values in $H^0(P; \mathbf{R})$. It can be shown that these are precisely the 2-cocycles and 2-coboundaries discussed above, and the component of $[\Omega_A]$ in this summand is just the obstruction $\tilde{\rho}$ to finding a lift of ρ to a Lie algebra homomorphism. We can now interpret $\tilde{\rho}$ geometrically: it is the locally constant function from P to $H_G^2(G; \mathbf{R})$ which assigns to each $p \in P$ the pullback of $[\Omega]$ under the orbit map $g \longmapsto A(g, p)$. The local constancy follows from the homotopy invariance of the induced map on cohomology.

Suppose that all the components of $[\Omega_A]$ are zero, so that there is a G-invariant

1-form φ on $G \times P$ with $d\varphi = \Omega_A$. For each $p \in P$, $\varphi(e, p)$ belongs to $T_e^*(G) \oplus T_p^*(P)$. Since $T_e(G) \approx \mathfrak{g}$, the first component of the map $p \longmapsto \varphi(e, p)$ may be considered as a mapping from P to $\mathfrak{g}^*$. Some computation shows that this mapping is a momentum function for the action ρ; the associated lift is the Lie algebra homomorphism from $\mathfrak{g}$ to $C^\infty(P; \mathbf{R})$ whose existence is guaranteed by the vanishing of the two obstructions. Furthermore, one may check that the momentum function is equivariant; here, G acts on P by the action A and on $\mathfrak{g}^*$ by the *coadjoint* (dual of the adjoint) representation.

The momentum function associated with a Lie algebra homomorphism lifting ρ turns out to give rise to a lagrangian submanifold. The action A determines the map $\widetilde{A} = (\pi_p, A)$ from $G \times P$ to $P \times P$. If we consider $P \times P$ as the symplectic manifold $(P, -\Omega) \times (P, \Omega)$, then Ω_A is just the pullback of this symplectic structure by A. If we identify $\mathfrak{g}^*$ with the left invariant 1-forms on G, then any map from P to $\mathfrak{g}^*$ can be considered as a G-equivariant map from $G \times P$ to T^*G. If μ is such a map, we have an embedding $(\mu, \widetilde{A}) = (\mu, \pi_p, A)$ from $G \times P$ to the symplectic manifold $(T^*G, \Omega_G) \times (P, -\Omega) \times (P, \Omega)$; a simple calculation shows that *μ is the momentum function associated with a Lie algebra homomorphism lifting ρ if and only if the embedding (M, A) is lagrangian.* Quantization theory (see Lecture 10) will provide us with a nice interpretation of this construction.

Lecture 5. Normal Forms

The basic normal form theorem in symplectic algebra states that every symplectic vector space (V, Ω) has a canonical basis, i.e. a basis $e_1, \ldots, e_n, f_1, \ldots, f_n$ such that

$$\Omega(e_i, e_j) = \Omega(f_i, f_j) = 0 \qquad \Omega(e_i, f_j) = -\Omega(f_j, e_i) = \delta_{ij}.$$

This result, proved at the beginning of Lecture 2, implies that (V_1, Ω_1) and (V_2, Ω_2) are symplectomorphic if and only if V_1 and V_2 have the same dimension, i.e. if and only if they are isomorphic as vector spaces.

The simplest extension of this result to manifolds is Darboux's theorem, which states that, if (P, Ω) is any symplectic manifold and p is any point of P, then there are coordinates $(x_1, \ldots, x_n, \xi_1, \ldots, \xi_n)$ on some neighborhood U of p such that $\Omega = \Sigma_{i=1}^n dx_i\, d\xi_i$ on U; in other words, two symplectic manifolds are *locally* symplectomorphic if and only if they have the same dimension. On the other hand, the global problem is virtually unsolved, except in dimension 2, where Moser has proven that compact surfaces (P_1, Ω_1) and (P_2, Ω_2) are symplectomorphic if and only if $X(P_1) = X(P_2)$ (i.e. P_1 and P_2 are diffeomorphic) and $\int_{P_1} \Omega_1 = \int_{P_2} \Omega_2$.

As Moser himself has noted, the method of proof of his theorem about surfaces can also be used to give a new proof of Darboux's theorem. We will now describe this method and then indicate how it can be applied in a very wide variety of situations.

Suppose that Ω_0 and Ω_1 are symplectic structures on P such that, for all $t \in [0, 1]$, the closed form $\Omega_t = \Omega_0 + t(\Omega_1 - \Omega_0)$ is also a symplectic structure. (This is always the case if Ω_0 and Ω_1 are sufficiently close, or if $\dim P = 2$ and Ω_0 and Ω_1 have the same sign.) Suppose, further, that Ω_0 and Ω_1 are cohomologous, so that $\Omega_1 - \Omega_0 = d\varphi$ for a 1-form φ (e.g. if P is a compact surface and $\int_P \Omega_0 = \int \Omega_1$). If we let ξ_t be the vector field $-\widetilde{\Omega}_t^{-1} \circ \varphi$, we have $L_{\xi_t} \Omega_t = d(\xi_t \lrcorner \Omega_t) = -d\varphi = \Omega_0 - \Omega_1$. Now if the time-dependent vector field $\{\xi_t\}$ can be integrated to a family $\{f_t\}$ of diffeomorphisms with $f_0 =$ identity and $df_t/dt = \xi_t \circ f_t$ (e.g. if P is compact), then we have

$$\frac{d}{dt}[f_t^* \Omega_t] = f_t^* \left[\frac{d\Omega_t}{dt} + L_{\xi_t} \Omega_t \right] = f_t^*[\Omega_1 - \Omega_0 + \Omega_0 - \Omega_1] = 0$$

so $f_t^* \Omega_t$ is constant, and $f_1^* \Omega_1 = f_0^* \Omega_0 = \Omega_0$; i.e. (P, Ω_0) and (P, Ω_1) are symplectomorphic.

To prove Darboux's theorem by this method, given $p \in P$ we first choose a symplectic basis for T_pP and use this to construct coordinates $(x_1, \ldots, x_n, \xi_1, \ldots, \xi_n)$ about p such that Ω and $\Sigma_{i=1}^n dx_i \wedge d\xi_i$ agree *at the point p*. Now let $\Omega_1 = \Sigma_{i=1}^n dx_i \wedge d\xi_i$ and $\Omega_0 = \Omega$. Since $\Omega_t = \Omega_0 = \Omega_1$ at the point p, and since the condition that a skew-symmetric form be a symplectic structure is *open*, and since the interval $[0, 1]$ is compact, it follows that there is a neighborhood U of p on which all the Ω_t's are symplectic. The only obstacle to completing the construction outlined above is that the field $\{\xi_t\}$ may not be integrable because U is not compact. To avoid this difficulty, we may add a closed form, if necessary, to the form φ such that $d\varphi = \Omega_1 - \Omega_0$, so that we have $\varphi(p) = 0$. Now $\xi_t(0) = 0$ for all t, so the flow $\{f_t\}$ leaves p fixed, and we may choose a neighborhood U of p such that f_t is an embedding of U into p for all t. For $t = 1$, we have $f_1^*(\Sigma dx_i \wedge d\xi_i) = \Omega$ on u, so f_1 is a symplectomorphism between the open subset U of (P, Ω) and the open subset $f_1(U)$ of $(\mathbf{R}^{2n}, \Sigma dx_i \wedge d\xi_i)$.

The essential step in our proof of Darboux's theorem was showing that, if two symplectic structures on P have the same value at a point $p \in P$, then their restrictions to sufficiently small neighborhoods of p are symplectomorphic. The same idea can be applied if we replace p by a submanifold of P: one obtains the following result.

EXTENSION THEOREM. *Let P be a manifold, $N \subseteq P$ a closed submanifold.*

(a) *Let Ω be a skew-symmetric bilinear form on the restricted tangent bundle $T_N P$ whose restriction to TN is a closed form on N. Then Ω extends to a closed form on a neighborhood of N in P. If Ω is a symplectic structure, so is its extension.*

(b) *Let Ω_0 and Ω_1 be symplectic structures on P whose restrictions to $T_N P$ are equal. Then there are neighborhoods U and V of N in P and a symplectomorphism f from (U, Ω_0) to (V, Ω_1) such that $f | N$ and $Tf | T_N P$ are the identity mappings.*

Both parts of the extension theorem rely on the following

RELATIVE POINCARÉ LEMMA. *Let $N \subseteq P$ be a closed submanifold, Ω a closed k-form on P whose pullback to N is zero. Then there is a $p - 1$-form φ on a neighborhood of N in P such that $d\varphi = \Omega$ and φ vanishes on N. If Ω vanishes on N, then φ can be chosen so that its first partial derivatives vanish on N.*

To prove the relative Poincaré lemma, one uses the homotopy operator associated with a deformation of a tubular neighborhood of N into N. The proof of the extension theorem is now left to the reader (see [W 2]). (We remark that the theorem is also true for nowhere vanishing 1-forms and volume forms.)

As an application of the extension theorem, we can classify symplectic neighborhoods of isotropic submanifolds. If $I \subseteq P$ is an isotropic submanifold, then $TI^\perp$ is a coisotropic submanifold of $T_I P$ and $TI^\perp/TI$ is a symplectic vector bundle over I which we call the *symplectic normal bundle* of I in P and denote by $SN(I, P)$. On the other hand, the quotient $T_I P/TI^\perp$ is naturally isomorphic to T^*I by the mapping $\alpha(x)(y + TI^\perp) = \Omega(x, y)$, so the exact sequence

$$0 \longrightarrow TI^\perp/TI \longrightarrow T_I P/TI \longrightarrow T_I P/TI^\perp \longrightarrow 0$$

becomes

$$0 \longrightarrow SN(I, P) \longrightarrow T_I P/TI \longrightarrow T^*I \longrightarrow 0.$$

In other words, the ordinary normal bundle of I in P, T_IP/TI, is isomorphic to the direct sum of T^*I and the symplectic normal bundle $SN(I, P)$. It follows that T_IP is isomorphic to $TI \oplus T^*I \oplus SN(I, P)$. Now $TI \oplus T^*I$ is a symplectic vector bundle as well, and in fact it is easy to verify that T_IP is isomorphic as a symplectic vector bundle to $TI \oplus T^*I \oplus SN(I, P)$. Applying part (b) of the extension theorem, we may conclude that, if I is embedded as an isotropic submanifold in (P_1, Ω_1) and (P_2, Ω_2), then the existence of neighborhoods U_i of I in P_i and a symplectomorphism $f\colon U_1 \longrightarrow U_2$ which is the identity on I is equivalent to the existence of a symplectic vector bundle isomorphism between $SN(I, P_1)$ and $SN(I, P_2)$. Next, we may show that every symplectic vector bundle E over I arises as the symplectic normal bundle of an isotropic embedding. In fact, we may let $P = T^*I \oplus E$. Identifying I with the zero section of this bundle, we have a natural isomorphism $T_IP \approx TI \oplus T^*I \oplus E$, and there is a natural symplectic structure on this bundle. The restriction of this to TI is zero, hence a closed form, so we may apply part (a) of the extension theorem to find a symplectic structure on a neighborhood of I in P for which E is the symplectic normal bundle. Summarizing our results, we have:

ISOTROPIC MANIFOLD THEOREM. *Let I be a manifold of dimension k. Then the extensions of I to a $2n$ dimensional symplectic manifold in which I is isotropic are classified, up to local symplectomorphism about I, by the isomorphism classes of $2(n - k)$ dimensional symplectic vector bundle over I.*

EXAMPLES. (i) A simple case is that of lagrangian submanifolds, i.e. $n = k$. In this case, the symplectic normal bundle reduces to zero, and we conclude that every symplectic manifold containing I as a lagrangian submanifold is, near I, symplectomorphic to a neighborhood of the zero section in T^*I. We shall see in Lectures 5 and 7 some consequences of this fact, which is proven in [**W 2**].

(ii) The case $k = 0$ is just Darboux's theorem.

(iii) If $k = 1$ (so $I = \mathbf{R}$ or S^1), then the symplectic normal bundle must be trivial, so P must be symplectomorphic near I to $T^*I \times \mathbf{R}^{2(n-1)}$.

(iv) The symplectic manifolds in which I is an isotropic submanifold of codimension $\dim I + 2$ are classified by the 2 dimensional symplectic bundles over I. But these are in 1-1 correspondence (by positive polarizations), with the complex line bundles over I, and these are in 1-1 correspondence with the elements of $H^2(I; \mathbf{Z})$.

A further refinement of the normal form theorems can be obtained if there is a group G acting on P. If G preserves symplectic structures Ω_0 and Ω_1 which agree along a submanifold N, and if G commutes with a deformation retraction of a tubular neighborhood of N into N, then the map such that $f^*\Omega_1 = \Omega_0$ can be constructed so that it commutes with the action of G. For instance, a symplectic action of a compact group near a fixed point is always symplectically conjugate to a symplectic linear action on $\mathbf{R}^n$. (This fact has been used by [**SEM**] in connection with Kirillov's integral for the Plancherel density.) The equivariant normal form theorem applied to an abelian group implies that any two isotropic foliations are locally symplectomorphic.

Lecture 6. Lagrangian Submanifolds and Families of Functions

A function S on a manifold X "generates" the lagrangian submanifold $dS(X)$. In this way, we can generate any lagrangian submanifold $L \subseteq T^*X$ locally if the projection $L \longrightarrow X$ is a diffeomorphism and globally if the pullback of ω_X to L is exact. In this lecture, we shall show how any lagrangian submanifold can be generated locally by a function on the product of X with a parameter space, and we shall discuss the corresponding global problem.

Locally, the construction goes as follows. Let $(x_1, \ldots, x_n)$ be coordinates on X, $(a_1, \ldots, a_k)$ some extra parameters, $S(x_1, \ldots, x_n, a_1, \ldots, a_k)$ a real valued function. We define the *critical set along the fibres*, Σ_S, by the equations

$$\partial S/\partial a_i = 0 \qquad i = 1, \ldots, k.$$

If these equations are independent, i.e. if the k by $n + k$ matrix

$$\begin{pmatrix} \dfrac{\partial^2 S}{\partial a_i \partial a_j} & \dfrac{\partial^2 S}{\partial a_i \partial x_j} \end{pmatrix}$$

has rank k, then Σ_S is a manifold of dimension n in $\mathbf{R}^{n+k}$. In this case, we say that S is nondegenerate or a *Morse family*. Now we define a mapping $\lambda_S\colon \Sigma_S \longrightarrow \mathbf{R}^{2n}$ by

$$\lambda_S(x_1, \ldots, x_n, a_1, \ldots, a_k) = \left(x_1, \ldots, x_n, \frac{\partial S}{\partial x_1}(x, a), \ldots, \frac{\partial S}{\partial x_k}(x, a)\right).$$

We will see that λ_S is a *lagrangian immersion* of Σ_S into $(\mathbf{R}^{2n}, \Sigma dx_i \wedge d\xi_i)$. This can be verified by direct calculation, but we prefer to use an approach which exhibits the symplectic nature of the construction.

Since S is a function on $\mathbf{R}^{n+k}$, the image L_S of dS is a lagrangian submanifold in $\mathbf{R}^{2(n+k)}$ (coordinates $(x_1, \ldots, x_n, a_1, \ldots, a_k, \xi_1, \ldots, \xi_n, \alpha_1, \ldots, \alpha_k)$) defined by the equations $\xi_i = \partial S/\partial x_i$, $\alpha_i = \partial S/\partial a_i$. In forming Σ_S, we are intersecting L_S with the coistropic submanifold $N \subseteq \mathbf{R}^{2(n+k)}$ defined by the equations $\alpha_1 = \cdots = \alpha_k = 0$. The subbundle $N^\perp$ is generated by the vector fields $\partial/\partial a_1, \ldots, \partial/\partial a_k$, so the quotient manifold $N/N^\perp$ is naturally identified with $\mathbf{R}^{2n}$ (coordinates $(x_1, \ldots, x_n, \xi_1, \ldots, \xi_n)$), and we see that $\lambda_S(\Sigma_S)$ is precisely the image of L_S under the reduction of $\mathbf{R}^{2(n+k)}$ to $\mathbf{R}^{2n}$. Finally, we may observe that the condition that S be nonsingular is equivalent to the condition that L_S be transversal to N, so we conclude from the theory of reductions (Lecture 3) that $\lambda_S(\Sigma_S)$ is an immersed lagrangian submanifold.

There is an invariant description of the construction above. If $B \xrightarrow{\pi} X$ is a differentiable submersion, the *conormal bundle to the fibres* is the subset $N_\pi \subseteq T^*B$ consisting of those cotangent vectors which annihilate the kernel of $T\pi$. By looking at π in local coordinates as above, we may see that N_π is coisotropic and that there is a natural map $N_\pi/N_\pi^\perp \xrightarrow{p} T^*X$ which is a local diffeomorphism (it is onto iff π is onto and 1-1 iff the fibres of π are connected). In fact, N_π may be identified with the pullback to B of T^*X.

Now let S be any real-valued function on B; we think of S as a *family of functions over* X. We call S a *Morse family* if $dS(X)$ is transverse to N_π. Then the reduction of $dS(X)$ to $N_\pi/N_\pi^\perp$ is an immersed lagrangian submanifold which we say is *generated by the family S.*

For example, let f be any real-valued function on $\mathbf{R}^n$ having 0 as a regular value, and define S on $\mathbf{R}^{n+1}$ by $S(x_1, \ldots, x_n, a) = a^3/3 + f(x_1, \ldots, x_n)a$. Σ_S is defined by the equation $0 = \partial S/\partial a = a^2 + f(x_1, \ldots, x_n)$; it is the *double* of the manifold with boundary in $\mathbf{R}^n$ defined by $f(x_1, \ldots, x_n) \leqslant 0$. The lagrangian immersion $\lambda_S\colon \Sigma_S \longrightarrow \mathbf{R}^{2n}$ is given by $\lambda_S(x_1, \ldots, x_n, a) = (x_1, \ldots, x_n, a\partial f/\partial x_1, \ldots, a\partial f/\partial x_n)$. In particular, if $f(x_1, \ldots, x_n) = x_1^2 + \cdots + x_n^2 - 1$, then Σ_S is the n-sphere whose equation is $x_1^2 + \cdots + x_n^2 + a^2 = 1$, and $\lambda_S(x_1, \ldots, x_n, a) = (x_1, \ldots, x_n, 2x_1a, \ldots, 2x_na)$. This lagrangian immersion has exactly one point of self intersection: $\lambda_S(0, 0, \ldots, -1) = \lambda_S(0, 0, \ldots, 1)$. If $n = 1$, the image of λ_S is a figure eight lying on its side in $\mathbf{R}^2$.

The next result shows that every lagrangian submanifold of T^*X is generated locally by a family over X, and it gives a sufficient condition for global generation. We denote by $VX \subseteq T^*X$ the "vertical" bundle tangent to the fibres of $T^*X \xrightarrow{\pi_X} X$.

THEOREM. *Let L be a lagrangian submanifold of T^*X such that:*

(i) *the restriction to L of ω_X is exact;*

(ii) *there is a lagrangian subbundle $\Lambda \subseteq T_L(T^*X)$ which is transversal to both V_LX and TL.*

Then L is generated by a family over X. In fact, one can take B to be a neighborhood of L in T^*X, $B \xrightarrow{\pi} X$ the restriction of π_X, and $\Sigma_S = L$.

To prove this theorem, we use the extension theorem in Lecture 5 to construct a symplectomorphism f from a tubular neighborhood U of L in T^*X onto a neighborhood V of the zero section Z in T^*L such that T_Lf maps Λ onto the vertical bundle V_ZL. Now $d(\omega_X - f^*\omega_L) = 0$ on U. Since ω_X is exact on L and $f^*\omega_L = 0$ on L, we can find a function S on U with $dS = \omega_X - f^*\omega_L$. Some calculation, using the fact that Λ is transversal to V_LX, shows that S is a Morse family when restricted to a sufficiently small neighborhood B of L, that Σ_S is exactly L, and that $\lambda_S\colon \Sigma_S \longrightarrow T^*X$ is the inclusion.

Condition (ii) of the theorem above is not necessary for L to be generated by a family. For instance, on $\mathbf{R}^3$, let $S(x, a_1, a_2) = -a_1^3/3 + xa_1 - a_2^3/3 + (1-x)a_2$; then Σ_S is the circle defined by $a_1^2 + a_2^2 = 1$ and $x = a_1^2$, while $\lambda_S(a_1^2, a_1, a_2) = (a_1^2, a_1 - a_2)$ maps Σ_S onto an upright figure eight. It is now easy to verify that there can be no continuous vector field along this "8" which is transversal to the vertical and the "8" itself.

It is not hard to see that condition (i) of the theorem above *is* necessary for L to be generated by a family. Furthermore, there is a necessary condition which is of a similar nature to (ii). At the end of this lecture, we will define the Maslov class of a pair of lagrangian subbundles in a symplectic vector bundle. It vanishes if the subbundles are homotopic to one another, in particular, if they admit a common lagrangian transversal. Then one can prove that, if L is generated by a family over X, the Maslov class of the pair $V_L X$ and TL is zero. The simplest example for which this condition is not satisfied is the circle $x^2 + \xi^2 = 1$ in $\mathbf{R}^2$.

Finally, there is the question of uniqueness–if two families over X generate the same lagrangian submanifold of T^*X, in what sense are they "the same"? To answer this question, we describe two operations on families which do not change the lagrangian submanifold being generated. First, if $B_i \xrightarrow{\pi_i} X$ is a submersion and S_i is a function on B_i, for $i = 1, 2$, we say that the families S_1 and S_2 are *diffeomorphic* if there is a diffeomorphism $g: B_1 \longrightarrow B_2$ such that $\pi_2 \circ g = \pi_1$ and $S_2 \circ g = S_1$. Diffeomorphic families obviously generate the same lagrangian submanifold. Second, if S is a family defined on $B \xrightarrow{\pi} X$, and k and l are nonnegative integers, we define the $k - l$ *suspension* of S as the function $S_{k,l}$ on $B \times \mathbf{R}^{k+l}$ given by $S_{k,l}(b, r_1, \ldots, r_k, s_1, \ldots, s_l) = S(b) + r_1^2 + \cdots + r_k^2 - s_1^2 - \cdots - s_l^2$. Since the conditions $\partial s_{k,l}/\partial r_i = 0$ and $\partial s_{k,l}/\partial s_i = 0$ force us back onto $B \times \{0, 0, \ldots, 0\}$, the lagrangian submanifold generated by $S_{k,l}$ is the same as that generated by S. Now we have the result:

EQUIVALENCE THEOREM. *Suppose that S_1 on $B_1 \xrightarrow{\pi_1} X$ and S_2 on $B_2 \xrightarrow{\pi_2} X$ generate the same lagrangian submanifold $L \subseteq T^*X$. Let $b_1 \in B_1$ and $b_2 \in B_2$ be points which correspond to the same point of L. Then there are neighborhoods U_i of b_i in B_i and nonnegative integers k_i, l_i such that the restriction of S_{1,k_1,l_1} to U_1 is diffeomorphic to the restriction of S_{2,k_2,l_2} to U_2.*

A proof of this theorem is outlined in **[WE 1]**. Details are given in **[D]** and **[HÖ]**. In short, it states that the germs of lagrangian submanifolds of T^*X are in 1-1 correspondence with the stable (with respect to suspension) diffeomorphism classes of germs of Morse families over X. Recent work (see **[AN 5]**), has been devoted to the classification of these objects when the allowable equivalences are expanded to include the diffeomorphisms of X.

The equivalence theorem suggests an approach to the question of existence of global generating families. Given a lagrangian submanifold $L \subseteq T^*X$, we can cover L by open sets $\{U_\alpha\}$ such that each $U_\alpha \cap U_\beta$ is connected, and each U_α is generated by a family $S_\alpha: U_\alpha \times \mathbf{R}^{k_\alpha} \longrightarrow \mathbf{R}$. Now the equivalence theorem implies that, on $U_\alpha \cap U_\beta$,

$$\operatorname{index}\left(\frac{\partial^2 S_\alpha}{\partial a_i \partial a_j}\right)_{1 \leq i, j \leq k_\alpha} - \operatorname{index}\left(\frac{\partial^2 S_\beta}{\partial a_i \partial a_j}\right)_{1 \leq i, j \leq k_\beta}$$

is a constant $C_{\alpha\beta}$. The $C_{\alpha\beta}$'s define a 1-dimensional Čech cocycle on L whose cohomology class $m_L \in H^1(L; \mathbf{Z})$ is an invariant of L called the *Maslov class* of L. (See **[MA]**, **[AN 2]**.) (It depends only on the lagrangian subbundles TL and $V_L X$ in $T_L(T^*X)$; see **[HÖ]**.) The vanishing of the class looks like a sufficient condition for the existence of a global generating family for a neighborhood of the 1-skeleton of L (with respect to some cell decomposition). To describe the higher obstructions, it might be useful to consider, not just

$$\operatorname{index}\left(\frac{\partial^2 S_\alpha}{\partial a_i \partial a_j}\right),$$

but the actual "negative bundle" E_α of the "fibre-hessian" of S_α. E_α is not really a vector bundle; since its fibres jump in dimension. Since the jumps of E_β are the same, though, it is conceivable that the difference $E_\alpha - E_\beta$ might be a well-defined element in K-theory. This speculation is supported by recent (unpublished) work of Lees, which indicates that the existence of a global generating family for L is an isotopy-invariant property, so that there should be *some* obstruction theory of a homotopic nature.

Lecture 7. Intersection Theory of Lagrangian Submanifolds

Let L_1 and L_2 be compact lagrangian submanifolds of (P, Ω) which are C^1 close together in the sense that there is a diffeomorphism $i: L_1 \longrightarrow L_2$ which is C^1 close, as a map into P, to the inclusion of L_1 in P. Then the intersection $L_1 \cap L_2$ is equal to the zero set of a closed 1-form on L_1, as the following argument shows. By the normal form theorem for lagrangian submanifolds, a neighborhood $\mathcal{U}$ of L in P is symplectomorphic to a neighborhood of the zero section in T^*L_1. We assume that L_2 is C^0 close enough to L_1 so that it lies in $\mathcal{U}$ – then we may as well assume that (P, Ω) is (T^*L_1, Ω_{L_1}), and that L_1 is the zero section. Now if the projection of L_2 on L_1 is C^1 close enough to the identity, it is a diffeomorphism, and L_2 must be the image of a 1-form on L_1, which is closed because L_2 is lagrangian. The zeros of this 1-form are evidently the points of $L_1 \cap L_2$.

A simple corollary of this observation is the following fixed point theorem.

THEOREM. *Let (P, Ω) be a compact symplectic manifold with $H^1(P; \mathbf{R}) = 0$. Then any symplectomorphism $f: P \longrightarrow P$ which is sufficiently C^1 close to the identity has at least two fixed points.*

PROOF. In $(P, \Omega) \times (P, -\Omega)$, let L_1 be the diagonal and L_2 the graph of f. If f is sufficiently C^1 close to the identity, there is a closed 1-form ω on L_1 whose zeros are the points of $L_1 \cap L_2$; these correspond to the fixed points of f. Now L_1 is (naturally) homeomorphic to P, so $H^1(L_1; \mathbf{R}) = 0$, and ω must be dS for some function S on L_1. Since S attains its maximum and minimum on L_1, $dS = \omega$ is zero at at least two points.

This result has some of the flavor of Poincaré's "last geometric theorem", though it is much simpler. Poincaré's theorem states that any area-preserving mapping of an annulus which "twists the bounding curves in opposite directions" has at least two fixed points. A first step toward generalizing our theorem in the direction of Poincaré's would be to replace the restriction "C^1 close" by "C^0 close", and then possibly by "isotopic". Toward this end, it would be interesting to develop further the intersection theory of lagrangian submanifolds, a subject which is intimately related to the study of self-intersections of lagrangian immersions.

A discussion of other possible generalizations of Poincaré's theorem is contained in a supplement by Arnol'd to the Russian translation of Poincaré's collected works **[AN 3]**.

We have seen that the intersection problem for lagrangian submanifolds leads to the differential-topological problem of classifying manifolds admitting a nowhere-zero closed 1-form. Tischler **[TI]** has shown that a manifold X admits such a form if and only if it fibres over a circle. The obstruction to such a fibration involves algebraic K-theory **[F]**.

One manifold which fibres over the circle is the torus $\mathbf{T}^n$. If we identify $T^*\mathbf{T}^n$ with $\mathbf{T}^n \times \mathbf{R}^n$, a 1-form without zeros on $\mathbf{T}^n$ may be considered as a mapping $\sigma\colon \mathbf{T}^n \longrightarrow \mathbf{R}^n\backslash\{0\}$. Recently, F. Laudenbach and A. Douady **[LB]** have shown that, if the 1-form is closed, then the map σ is necessarily homotopic to a constant. It is possible to find an $n-1$ torus $\Sigma \subset \mathbf{T}^n$, a neighborhood U of Σ, and a nowhere zero closed 1-form φ on U such that

(i) the map $\sigma\colon U \longrightarrow \mathbf{R}^n\backslash\{0\}$ is not homotopic to a constant;

(ii) φ extends to $\mathbf{T}^n$ as a closed 1-form;

(iii) φ extends to $\mathbf{T}^n$ as a nowhere-zero 1-form.

Due to (i), φ does not extend to $\mathbf{T}^n$ as a nowhere-zero closed 1-form. It would be interesting to have a "symplectic" interpretation of this example.

The intersection theory of lagrangian submanifolds can also be used to obtain results on boundary value problems for hamiltonian flows. If we have an action $\mathbf{R} \times P \xrightarrow{A} P$ with a momentum (hamiltonian) function μ, we may consider, as in Lecture 4, the lagrangian embedding $(\mu, \widetilde{A})$ from $\mathbf{R} \times P$ to $\mathbf{R} \times \mathbf{R}^* \times P \times P$ given by $(t, p) \longrightarrow (t, \mu(p), p, A(t, p))$. Any lagrangian submanifold $\Lambda \subseteq \mathbf{R} \times \mathbf{R}^* \times P \times P$ is said to define a *canonical boundary value problem* for actions of $\mathbf{R}$ on P; the *solutions* of this problem are the points (t, p) in $(\mu, \widetilde{A})^{-1}(\Lambda)$. Lagrangian intersection theory may be used to study the bifurcation of solutions under perturbations of A or Λ.

For instance, if $\Lambda = C \times \Delta$, where C is a curve in $\mathbf{R} \times \mathbf{R}^*$ and Δ is the diagonal in $P \times P$, then the solutions of Λ correspond to periodic points (t, p) of A which satisfy the "period-energy" relation C.

Applications of this idea may be found in **[AN 1]**, **[BA]**, **[W 3]**, **[W 4]**.

I wonder whether there are interesting applications to groups other than $\mathbf{R}$. (See the end of Lecture 4.)

Lecture 8. Quantization on Cotangent Bundles

In physics, the term "quantization" refers to the process of passing from a classical-mechanical description of a physical system to a quantum-mechanical description of the same system. The mathematical model for a (conservative) classical system is a symplectic manifold together with a group of symplectomorphisms (symmetries and/or time-translations); for a quantum system the symplectic manifold is replaced by a complex Hilbert space and the symplectomorphisms by unitary operators.

The simplest mathematical formulation of the quantization problem would be, therefore, to find a functor from the "classical" category

$$\mathcal{C} = (\text{symplectic manifolds, symplectomorphisms})$$

to the "quantum" category

$$\mathcal{Q} = (\text{complex Hilbert spaces, unitary operators}).$$

This functor should satisfy some extra conditions suggested by physics or mathematics. It was shown by Van Hove **[HV]** that, if one imposes some mild conditions suggested by the Schrodinger quantization used in physics, no such functor exists.

To reformulate the quantization problem so that it has a chance for solution, we require only a *functorial relation* from $\mathcal{C}$ to $\mathcal{Q}$; this is to be a subcategory of $\mathcal{C} \times \mathcal{Q}$, not necessarily the graph of a functor. One way to obtain a functorial relation is to construct a pair of functors $\mathcal{C} \longleftarrow \mathcal{D} \longrightarrow \mathcal{Q}$, where $\mathcal{D}$ is some auxiliary category. Such a pair, perhaps satisfying some extra conditions, may be called a *quantization procedure.* (We do not wish to suggest that this is the only possible mathematical formulation of the quantization problem. Quite different ones are given, for example, in **[BE]** and **[MA]**. See also **[H]** for a general discussion of quantization.)

The basic quantization procedure is the cotangent bundle–1/2-density relation derived from the physicists' treatment of point transformations. The auxiliary category for this procedure is

$$\mathcal{M} = (C^\infty \text{ manifolds, } C^\infty \text{ diffeomorphisms}).$$

From $\mathcal{M}$ to $\mathcal{C}$ we have the functor T^* which associates to each manifold X its cotangent bundle and to each diffeomorphism $f\colon X \longrightarrow Y$ its natural lift $T^*f\colon T^*X \longrightarrow T^*Y$. To define a functor from $\mathcal{M}$ to $\mathcal{Q}$, we begin with the bundle $|X|^{1/2}$ of 1/2-densities on X. (The fibre at $x \in X$ of the bundle of α-densities on X consists of those complex-valued functions δ on the set of bases of T_xX such that, for each matrix $(b_{ij}) \in GL(n, \mathbf{R})$ and each frame

$\{e_i\}$, we have $\delta\{\Sigma_j b_{ij} e_j\} = |\det b_{ij}|^\alpha \delta\{e_i\}$.) If γ_1 and γ_2 are in the space $\mathcal{D}(|X|^{1/2})$ of compactly supported C^∞ sections of $|X|^{1/2}$, the pointwise product $\gamma_1\bar{\gamma}_2$ is a section of $|X|^1$ which can be integrated over X to give a complex number. The pairing thus defined makes the space $\mathcal{D}(|X|^{1/2})$ into a pre-Hilbert space; the completion of $\mathcal{D}(|X|^{1/2})$ is denoted by $H(X)$; its elements are called L^2 1/2-densities on X. (If X has a distinguished volume element, $H(X)$ is naturally isomorphic to the space of L^2 functions on X with respect to this volume element, but the space $H(X)$ itself is independent of the choice of volume element; for a measure.theoretic definition of $H(X)$, see **[MA]**.) Since $H(X)$ is defined in a completely canonical way, it follows that every diffeomorphism $X \xrightarrow{f} Y$ induces a unitary operator $H(X) \xrightarrow{H(f)} H(Y)$, so that H is a functor from $\mathcal{M}$ to $\mathcal{Q}$.

The quantization procedure $\mathcal{C} \xleftarrow{T^*} \mathcal{M} \xrightarrow{H} \mathcal{Q}$ plays two important roles in the general theory. First of all, one may require as an axiom that it be contained in any acceptable quantization procedure. (It is this axiom which Van Hove showed to be incompatible with the existence of a quantization functor.) Second, the cotangent bundle–1/2-density relation suggests some further analogies between structures in $\mathcal{C}$ and $\mathcal{Q}$ (see **[SL 1]**); one may require that a quantization procedure be compatible with these analogies. For example, if X and Y are manifolds, then $T^*(X \times Y) \cong T^*X \times T^*Y$, while $H(X \times Y) \cong H(X) \hat{\otimes} H(Y)$ (topological tensor product). This suggests that any acceptable functorial relation in $\mathcal{C} \times \mathcal{Q}$ which contains (P_1, E_1) and (P_2, E_2) should also contain $(P_1 \times P_2, E_1 \hat{\otimes} E_2)$.

Quantization procedures in physics often involve associating infinitesimal unitary transformations (i.e. skew-adjoint operators) on $H(X)$ with functions (thought of as the generators of infinitesimal symplectomorphisms) on T^*X. For instance, a function S on X pulled back to T^*X is often "quantized" by the operation of multiplication by iS on $H(X)$. The hamiltonian vector field of $S(q_1, \ldots, q_n)$ is $(\partial s/\partial q_1)(\partial/\partial p_1) + \cdots + (\partial s/\partial q_n)(\partial/\partial p_n)$, so the transformation obtained by exponentiating its hamiltonian vector field translates the fibre T^*_xX by the covector $dS(x)$. This suggests that we "quantize" this translation symplectomorphism by the operator of multiplication by e^{iS}. A problem arises here, since the function S is determined only up to an additive constant by the symplectomorphism of translation by dS; thus, the operator e^{iS} would be determined only up to a multiplicative constant.

It appears, then, that quantization should involve a category whose morphisms take into account the possible arbitrary constants in hamiltonian functions. In particular, the automorphism group associated with (P, Ω) should be an extension by S^1 of the group of symplectomorphisms of (P, Ω); the Lie algebra by the automorphism group should be the extension of $\mathcal{X}(P, \Omega)$ by $H^0(P; \mathbf{R})$ given by $C^\infty(P)$ with the Poisson bracket operation.

The problem of constructing such a category was solved in a special case by Van Hove **[HV]** and in general by Kostant **[KO]** and Souriau **[SR]**; we now describe their solution. If (P, Ω) is any symplectic manifold, a *prequantization* of (P, Ω) is defined by Kostant to be a complex line bundle together with a connection whose curvature form is Ω. In terms of principal bundles (Souriau's viewpoint), a prequantization of (P, Ω) consists of

(i) a manifold Q with a free S^1 $(= \mathbf{R}/2\pi\mathbf{Z})$ action generated by a vector field ξ;

(ii) a projection $Q \xrightarrow{\psi} P$ which identifies P with the orbit space Q/S^1;

(iii) a 1-form ω on Q such that $\xi \lrcorner \omega \equiv 1$ and $d\omega = \psi^*\Omega$.

The structures on Q are all determined by the 1-form ω. In fact, ξ is the unique vector field such that $\xi \lrcorner \omega \equiv 1$ and $\xi \lrcorner d\omega = 0$. Any manifold Q with a nowhere vanishing 1-form ω such that:

(i) $d\omega$, restricted to the subbundle of TQ annihilated by ω, is a symplectic structure,

(ii) the vector field ξ on Q such that $\xi \lrcorner \omega \equiv 1$ and $\xi \lrcorner d\omega = 0$ (well defined by (i)) has all its orbits periodic with period 2π,

is called a *regular contact manifold*; the quotient of Q by the orbits of ξ is a symplectic manifold. The regular contact manifolds form a category $\mathcal{P}$, the morphisms being diffeomorphisms which preserve the 1-forms; passing from Q to its quotient by orbits of ξ defines a functor σ from $\mathcal{P}$ to $\mathcal{C}$; a prequantization of (P, Ω) is essentially an object in $\sigma^{-1}(P, \Omega)$.

To determine the automorphism group of a prequantization, we observe that any contact automorphism f of (Q, ω) which projects to the identity on $\sigma(Q, \omega)$ must leave the vector field ξ and the subbundle $\ker \omega$ invariant; it follows that f rotates the fibres of Q by an amount which is locally constant on Q. We have, therefore, an exact sequence, writing (P, Ω) for $\sigma(Q, \omega)$: $0 \longrightarrow H^0(P; S^1) \longrightarrow \mathrm{Autom}(Q, \omega) \longrightarrow \mathrm{Sympl}(P, \Omega)$. To see that the corresponding sequence of Lie algebras is just that of Lecture 4, we may represent each $S \in C^\infty(P)$ by the sum of the horizontal lift of ξ_S to Q and the vertical vector field $S\xi$. One may check that this representation gives an isomorphism of $C^\infty(P)$ onto the Lie algebra of vector fields on Q which leave ω invariant.

The classification of prequantizations of (P, Ω) has a simple algebraic description. (P, Ω) has a prequantization if and only if Ω belongs to the image of $H^2(P; \mathbf{Z})$ in $H^2(P; \mathbf{R})$; i.e. if and only if the integral of Ω over every integer 2-cycle is an integer. The set of prequantizations, if nonempty, is an affine space of the character group $\mathrm{Hom}(\pi_1(P), S^1) = H^1(P; S^1)$. (See **[KO]**.)

If a Lie group G acts on (P, Ω), we may wish to lift the action to some prequantization (Q, ω). At the Lie algebra level, the obstructions to doing this are exactly the cohomology classes described in Lecture 4. If the obstructions vanish, and G is connected and simply connected, the action lifts. For general G, there may be further obstructions.

There is a "natural" prequantization for the subcategory of $\mathcal{C}$ consisting of cotangent bundles. Let $d\theta$ be the constant 1-form on S^1 $(\int_{S^1} d\theta = 2\pi)$. If $(P, \Omega) = (T^*X, \Omega_X)$, we let $Q = T^*X \times S^1$, $\omega = -\omega_X + d\theta$ (both forms pulled back to Q by the Cartesian projections). If g is a diffeomorphism from X to Y and T^*g the associated symplectomorphism from T^*X to T^*Y, then $(T^*g)^*(\omega_Y) = \omega_X$. It follows that the mapping $T^*g \times$ identity is a contact morphism from $T^*X \times S^1$ to $T^*Y \times S^1$.

We may summarize the results up to now by the following diagram of functors; the dotted line represents the one just constructed.

$$\begin{array}{ccccc} \mathcal{P} & \nwarrow & & & \\ \sigma \downarrow & & \ddots & & \\ \mathcal{C} & \xleftarrow{\;T\;} & \mathcal{M} & \xrightarrow{\;H\;} & \mathcal{Q}. \end{array}$$

In prequantizing symplectomorphisms of the type T^*g, the only special property which we used was the fact that they preserve canonical 1-forms. It is natural, then, to look for other symplectomorphisms with this property. However, we have the following proposition.

PROPOSITION. *If $T^*X \xrightarrow{f} T^*Y$ is a diffeomorphism such that $f^*\omega_Y = \omega_X$, then $f = T^*g$ for some diffeomorphism $X \xrightarrow{g} Y$.*

To prove this proposition, we first note that the zero set of ω_X is the zero section of T^*X; the same is true for Y, so f induces a diffeomorphism between the zero sections. Identifying these sections with X and Y, we have our $X \xrightarrow{g} Y$. Next, we use the vector field η_X defined by $\eta_X \lrcorner \Omega_X = \omega_X$; g maps η_X into η_Y. In local coordinates, $\Omega_X = \Sigma dq_i \wedge dp_i$, $\omega_X = \Sigma p_i dq_i$, so $\eta_X = -\Sigma p_i(\partial/\partial p_i)$. The group generated by η_X consists of multiplication by positive scalars; each integral curve is a ray which tends to a definite point of the zero section. It follows that f maps T_x^*X onto $T_{g(x)}^*Y$ for each $x \in X$ and takes zero vectors to zero vectors. The composition $(T^*g)^{-1} \circ f$ leaves invariant each fibre of T^*X, as well as the zero section; one may check (see [W 2]) that the only symplectomorphism with this property is the identity, so $f = T^*g$, and the proof is complete.

The proof above depends in a crucial way upon the presence of the zero sections. If we replace T^*X by $\dot{T}^*X$ (zero section deleted), then there are many new symplectomorphisms $\dot{T}^*X \xrightarrow{k} \dot{T}^*Y$ for which $k^*\omega_Y = \omega_X$. In fact, the proof above shows that these k are exactly the *homogeneous* symplectomorphisms, i.e. those which commute with multiplication by positive real numbers. There are plenty of these: if h: $\dot{T}^*X \longrightarrow \mathbf{R}$ is positively homogeneous of degree 1, then the hamiltonian flow generated by h consists of homogeneous symplectomorphisms.

A quantization procedure for homogeneous symplectomorphisms between "punctured" cotangent bundles is provided by the theory of Fourier integral operators [D],[HÖ]. The infinitesimal version of this theory has been in existence somewhat longer – it is the theory of pseudo-differential operators. If X is compact, given any $h \in C^\infty(T^*X)$, homogeneous of degree 1, there is an unbounded skew-Hermitian operator on $H(X)$, determined modulo the bounded operators, whose "principal symbol" is h. If h is *linear* on fibres, it comes from a vector field on X, and the corresponding operator is just Lie derivative by this vector field. The principal symbol correspondence is compatible with the Lie algebra structures in $\underset{\sim}{C}$ and $\underset{\sim}{Q}$.

In the theory of Fourier integral operators, there is associated to each homogeneous symplectomorphism C from $\dot{T}^*X$ to $\dot{T}^*Y$ a class I_C of bounded operators from $H(X)$ to $H(Y)$. The operators in I_C are Fredholm with the same index; if the index is zero, I_C contains unitary operators. (See [W 5]; it is unknown whether the index is always zero.)

We shall say more about Fourier integral operators in Lecture 10.

Lecture 9. Quantization and Polarizations

We will describe in this lecture the so-called "geometric quantization procedure" of Segal, Kirillov, Kostant, and Souriau.

Let (Q, ω) be a regular contact manifold. We denote the automorphism of (Q, ω) associated with $\theta \in S^1$ by θ_Q. Let $\mathcal{D}(Q, \omega)$ consist of the compactly supported C^∞ functions $\gamma: Q \longrightarrow \mathbf{C}$ which satisfy the equivariance law $\gamma(\theta_Q q) = e^{-i\theta}\gamma(q)$. Then $\mathcal{D}(Q, \omega)$ is a pre-Hilbert space with respect to the inner product

$$\langle \gamma_1, \gamma_2 \rangle = \frac{1}{2\pi} \int_Q \gamma_1(q)\overline{\gamma}_2(q)\omega \wedge (d\omega)^n;$$

we denote its completion by $K(Q, \omega)$. It is clear that any isomorphism from (Q_1, ω_1) to (Q_2, ω_2) induces a unitary operator from $K(Q_1, \omega_1)$ to $K(Q_2, \omega_2)$, so K is a functor K from $\mathcal{P}$ to $\mathcal{Q}$.

REMARK. This description of the functor K is due to Souriau **[SR]**. In Kostant's formulation **[KO]**, one considers Q as the principal S^1 bundle over P, with a connection whose curvature form is Ω. $K(Q, \omega)$ is then the space of L^2 sections of the associated complex line bundle, which we will denote by $B(Q, \omega)$.

The pair $\mathcal{C} \overset{\sigma}{\longleftarrow} \mathcal{P} \overset{K}{\longrightarrow} \mathcal{Q}$ fails as a quantization procedure because it is not consistent with the cotangent bundle – 1/2-density relation. To see this, we consider the case $Q = T^*X \times S^1$, $\omega = -\omega_X + d\theta$. $K(Q, \omega)$ consists of those L^2 functions on Q of the form $\gamma(x, \xi, \theta) = e^{-i\theta}\gamma(x, \xi, 0)$; so $K(Q, \omega)$ may be identified with the L^2 functions on T^*X. But (Q, ω) is a prequantization of (T^*X, Ω_X), and the cotangent bundle – 1/2-density relation associates that symplectic manifold with the space $H(X)$ of L^2 1/2-densities on X. In a sense, then, the space $K(Q, \omega)$ is "too big".

The space $K(Q, \omega)$ also violates another principle derived from physical considerations. Suppose that $\mathcal{C} \overset{A}{\longleftarrow} \mathcal{D} \overset{B}{\longrightarrow} \mathcal{Q}$ is a quantization procedure, and that G is a *group* of automorphisms of some object U of $\mathcal{D}$. Then $A(G)$ gives a symplectic action of G on $A(U)$, while $B(G)$ gives a unitary representation of G on $B(U)$. Comparing the notions of "indecomposibility" of systems in classical and quantum mechanics (see **[SL 1]**), we are led to postulate that *transitivity* (in some sense, perhaps topological) of the action $A(G)$ should be equivalent to *irreducibility* of the representation $B(G)$. If we let $X = \mathbf{R}$, for example, with G the group of automorphisms of $Q = \mathbf{R} \times \mathbf{R}^* \times S^1$ which project to translation on $\mathbf{R} \times \mathbf{R}^*$, then $\sigma(G)$ acts transitively on $(\mathbf{R} \times \mathbf{R}^*, \Omega_{\mathbf{R}})$ but the representation $K(G)$ on the L^2 1/2-densities on $\mathbf{R} \times \mathbf{R}^*$ is not irreducible.

To see how we might "cut down" the representation $K(Q, \omega)$, we look again at the case $Q = T^*X \times S^1$ and see how the space $H(X)$ of 1/2-densities on X can be recovered geometrically from Q. Recall that we identified $K(Q, \omega)$ with the space of L^2 functions on T^*X; if we momentarily forget the distinction between functions and 1/2-densities, then we may think of the elements of $H(X)$ as functions on T^*X which are constant along the fibres of the projection $T^*X \longrightarrow X$. Of course, these functions are no longer in L^2 on T^*X.

The following abstraction of the idea of using functions constant along the fibres was developed by Segal, Kirillov, Kostant, and Souriau. Given a symplectic manifold (P, Ω) and its prequantization (Q, ω), with projection $\psi: Q \longrightarrow P$, we add one more piece of structure, a *polarization* F of P (see Lecture 2). For the moment, we will assume that F is a real polarization, so that it is the tangent bundle to a foliation $\mathcal{F}$ of P. Let $\hat{F}$ be the horizontal lift of F to Q; i.e. for $q \in Q$, $\hat{F}_q$ is the unique subspace of T_qQ which is annihilated by ω and which projects isomorphically to $F_{\psi(q)}$. The fact that F is a *lagrangian* subbundle implies that, on the inverse image $\psi^{-1}(L)$ of each leaf of L, the form ω is closed. It follows that the lift $\hat{F}$ is again integrable and that the leaves of the resulting foliation $\hat{\mathcal{F}}$ of Q are coverings of the leaves of $\mathcal{F}$.

For each leaf L of $\mathcal{F}$, we now define the complex vector space $B^L(Q, \omega)$ to consist of those functions $\gamma: \psi^{-1}(L) \longrightarrow \mathbf{C}$ which, in addition to satisfying the equivariance condition $\gamma(\theta_Q q) = e^{-i\theta}\gamma(q)$, are constant on each leaf of $\hat{\mathcal{F}}$. In the Kostant picture, these are just the sections of B over L which are *covariant constant* with respect to the given connection; such sections always exist locally because the curvature form Ω pulls back to zero on L.

Since the leaf L is connected, an element of $B^L(Q, \omega)$ is determined by its value at any point of $\psi^{-1}(L)$, so $B^L(Q, \omega)$ is at most 1-dimensional. We will call the leaf L *quantizable* if $B^L(Q, \omega)$ does not reduce to zero. This will be the case whenever the leaves of $\hat{\mathcal{F}}$ in $\psi^{-1}(L)$ project diffeomorphically onto L, or equivalently, when the holonomy of the locally flat bundle $B(Q, \omega)$ over L is trivial. In particular, a simply connected leaf is always quantizable. When L is quantizable, we give a hermitian structure to $B^L(Q, \omega)$ by defining $\langle \gamma_1, \gamma_2 \rangle$ to be $\gamma_1(q)\overline{\gamma_2(q)}$, for any q in $\psi^{-1}(L)$. (It does not depend upon the choice of q, of course.)

We remark for use in Lecture 10 that the definition of $B^L(Q, \omega)$ and the notion of quantizability really depend only on the lagrangian submanifold L and not on the fact that L is a leaf of a foliation.

In the special case $Q = T^*X \times S^1$, $\omega = -\omega_X + d\theta$, F the tangent bundle to the fibres of $T^*X \longrightarrow X$, the leaf space $T^*X/\mathcal{F}$ is naturally identified with X, all leaves are quantizable, and each space $B^L(G, \omega)$ may be identified with $\mathbf{C}$ if we identify an element of $B^L(Q, \omega)$ with its value at the zero section of T^*X and the zero element of S^1. This suggests the following definitions.

Definition. A triple (Q, ω, F), where (Q, ω) is an object in $\mathcal{P}$ and F is a real polarization of $(P, \omega) = \sigma(Q, \omega)$, is called a *geometric quantum system* if the leaf space $P/\mathcal{F}$ is a manifold in the induced structure (i.e. if there is no recurrence in the leaves of $\mathcal{F}$) and if the set $X(Q, \omega, F)$ of quantizable leaves is a submanifold of $P/\mathcal{F}$.

The geometric quantum systems form a category $\mathcal{G}$ with respect to the obvious class of equivalences, and forgetting the polarizations gives a functor $\tau: \mathcal{G} \longrightarrow \mathcal{P}$. We will now define a functor $\Gamma: \mathcal{G} \longrightarrow \mathcal{Q}$ so that the pair $\mathcal{C} \xleftarrow{\sigma\cdot\tau} \mathcal{G} \xrightarrow{\Gamma} \mathcal{Q}$ becomes a quantization procedure.

Let (Q, ω, F) be a geometric quantum system, $(P, \Omega) = \sigma(Q, \omega)$, and write X for $X(Q, \omega, F)$. The 1-dimensional spaces $B^L(Q, \omega)$ for each quantizable leaf $L \in X$ fit together to form hermitian line bundle $E(Q, \omega, F)$ over X. Let $\mathcal{D}(Q, \omega, F)$ consist of the compactly supported C^∞ sections of the tensor product $E(Q, \omega, F) \otimes |X|^{1/2}$. The pointwise product $\gamma_1\overline{\gamma}_2$ of two elements of $\mathcal{D}(Q, \omega, F)$ may be considered as a section of $|X|^{1/2}$, so integration over X gives $\mathcal{D}(Q, \omega, F)$ a pre-Hilbert structure. The completion of $\mathcal{D}(Q, \omega, F)$ is a Hilbert space which we denote by $\Gamma(Q, \omega, F)$. Since the construction of $\Gamma(Q, \omega, F)$ was entirely canonical, a G-equivalence f from (Q_1, ω_1, F_1) to (Q_2, ω_2, F_2) induces a unitary transformation Γf from $\Gamma(Q_1, \omega_1, f_1)$ to $\Gamma(Q_2, \omega_2, f_2)$, and our functor Γ is defined. Furthermore, putting together our earlier remarks in this section, one may check that, if $Q = T^*X \times S^1$, $\omega = -\omega_X + d\theta$, and F is the polarization by the fibres, then $\Gamma(Q, \omega, F)$ is naturally isomorphic to $H(X)$, so the pair $\mathcal{C} \xleftarrow{\sigma \circ T} \mathcal{G} \xrightarrow{\Gamma} \mathcal{Q}$ is a reasonable quantization procedure. We will call it the SKKS procedure (for Segal, Kirillov, Kostant, and Souriau).

Of course, the importance of the SKKS procedure lies in the fact that it is more general than the cotangent bundle – 1/2-density procedure. The original application by Kirillov [KI 1] of this quantization procedure, was to the representation theory of nilpotent Lie groups. If G is a simply connected nilpotent Lie group, and P is an orbit in $\mathfrak{g}^*$ under the coadjoint representation, then:

- P has a natural G-invariant symplectic structure Ω;
- (P, Ω) has a unique prequantization (Q, ω);
- the action of G lifts to an action on (Q, ω);
- (P, Ω) admits a G-invariant real polarization F;
- P/F is a manifold, and all the leaves of P/F are quantizable, so (Q, ω, F) is a geometric quantum system on which G acts;
- applying the functor Γ to the action of G, we obtain a unitary representation of G on $\Gamma(Q, \omega, F)$;
- this unitary representation is irreducible;
- the representations arising from different choices of F (on a fixed orbit) are unitarily equivalent;
- every irreducible unitary representation of G arises, via the construction just described, from a unique orbit of G in $\mathfrak{g}^*$.

Kirillov originally described his construction of the representations associated with orbits in the language of induced representations. The differential-geometric content of the construction was later elucidated by Kostant and Souriau.

It is natural to try to extend Kirillov's results to other classes of Lie groups. It turns out that the SKKS procedure works quite well for solvable groups (see [AU], [BEN]). For semisimple groups, however, the situation is more complicated. First of all, Rothschild and Wolf [R] found that, for a certain coadjoint orbit of the exceptional Lie group G_2, two invariant real polarizations produced inequivalent representations. There is a "pairing" procedure, due to Kostant and Sternberg (see [BL 1]) which is meant to produce intertwining operators between the representations coming from different polarizations, but it does not work in the case studied by Rothschild and Wolf. Blattner **[BL 2]** has found the underlying geometric reason for the difficulty.

Geometric quantization theory becomes more powerful if we allow the use of nonreal polarizations. To define the Hilbert space $\Gamma(Q, \omega, F)$ when F is totally complex, say, we no longer have the leaves of the foliation F to aid us. Instead of requiring that a function on Q be constant along the leaves of a foliation, we must say that it is annihilated by certain (complex-valued) vector fields. This amounts to giving a holomorphic structure to the bundle $B(Q, \omega)$ and requiring that the sections be holomorphic.

The simplest application of the SKKS procedure with a complex polarization is to the case where $(P, \Omega) = (\mathbf{R}^2, dx \wedge d\xi)$ and G is the group of proper euclidean motions (for the metric $dx^2 + d\xi^2$). The group G leaves no real polarization invariant, but it does leave invariant the usual Kähler structure on $\mathbf{R}^2$ $(J\partial/\partial x = \partial/\partial\xi)$. G itself does not lift to the prequantization $(\mathbf{R}^2 \times S^1, -\xi\, dx + d\theta)$, but a 1-dimensional extension $\widetilde{G}$ does. The resulting representation of $\widetilde{G}$ on a certain space of entire functions on $\mathbf{C}$ is called the Fock-Segal-Bargmann representation (see **[BE]**). Applying the SKKS procedure with complex polarization to compact Lie groups acting on compact Kähler manifolds, Kostant obtained the Borel-Weil-Bott description of the irreducible representation spaces as sections of holomorphic vector bundles.

For general semisimple groups, there may be coadjoint orbits which admit no invariant polarizations to which the SKKS procedure may be applied. The orbit method remains a useful heuristic tool in these cases. Using analytic methods, one can sometimes associate representations to orbits by using noninvariant polarizations **[BL 1]**, by passing to a limit from orbits which do admit invariant polarizations **[O]**, or by using pseudo-differential operators **[AK]**. The orbit method also suggests a conjecture for the decomposition into irreducible representation of the action of a group G on the 1/2-densities on a homogeneous space G/H. This conjecture, due to Shubov **[SH]**, involves decomposing the cotangent bundle $T^*(G/H)$ into G orbits and then applying the reduction procedure (Lecture 3) to obtain coadjoint orbits. Shubov has verified his conjecture, which is a geometrization of Kirillov's result for nilpotent G, in the case of $SL(n; \mathbf{R})$ acting on $\mathbf{R}^n \setminus \{0\}$.

Lecture 10. Quantizing Lagrangian Submanifolds and Subspaces, Construction of the Maslov Bundle

Quantizing lagrangian submanifolds is more general than quantizing symplectomorphisms. If (P_i, Ω_i), $i = 1, 2$, are symplectic manifolds, and H_i is a Hilbert space associated with (P_i, Ω_i) by some quantization procedure, then the cotangent bundle – 1/2-density relation and the SKKS theory suggest that the topological tensor product $H = H_1^* \hat{\otimes} H_2$ be associated with the product $(P, \Omega) = (P_1, -\Omega_1) \times (P_2, \Omega_2)$. H may be considered as a dense subspace of the space $\mathrm{Op}(H_1, H_2)$ of operators from H_1 to H_2 – it consists of the so-called *Hilbert-Schmidt* operators. Now if $f\colon P_1 \longrightarrow P_2$ is a symplectomorphism, its graph is a lagrangian submanifold of (P, Ω); the quantization procedure should associate to f an element (or elements) of $\mathrm{Op}(H_1, H_2)$. (Since the operator is unitary, it cannot belong to H.)

These observations, together with the formal properties of lagrangian submanifolds discussed in Lecture 3, suggest the following generalized quantization problem: Given a quantization procedure $\mathcal{C} \xleftarrow{A} \mathcal{D} \xrightarrow{B} \mathcal{Q}$, find for each object D in $\mathcal{D}$ a relation between the set $L(A(D))$ of lagrangian submanifolds in $A(D)$ and some linear space $V(D)$ in which $B(D)$ is embedded as a dense subspace. The relations for various D's should be compatible with the quantization of $\mathcal{D}$-morphisms and the application of symplectomorphisms to lagrangian submanifolds.

The reader is invited to interpret, in this context, the lagrangian submanifold $(M, A)(G \times P)$ defined in the last paragraph of Lecture 4.

We turn now to the quantization of lagrangian submanifolds of cotangent bundles.

Many different lines of argument suggest that one should associate the lagrangian submanifold $dS(X) \subseteq T^*X$ with the function e^{iS} on X. Here we give an argument based on geometric quantization theory, using the quantization $(Q, \omega, F) = ((T^*X \times S^1), -\omega_X + d\theta$, fibres). All the leaves are quantizable, so the sections of $E(Q, \omega, F)$ over X may be identified with the functions on $T^*X \times S$ of the form $\varphi(x, \xi, \theta) = e^{-i\theta}a(x)$, where a is a function X.

We now ask the question: What condition on $a(x)$ will assure that the function $e^{-i\theta}a(x)$ is constant along the horizontal lifts of $L = dS(X)$ to $T^*X \times S^1$? L is defined by the equations (in local coordinates) $\xi_i = \partial S/\partial x_i$; a lift is given by an equation $\theta = u(x)$, where u is a function from X to S^1. Horizontality of the lift is given by the vanishing of the 1-form $-\omega_X + d\theta$, so we must have

$$0 = -\sum \xi_i dx_i + d\theta = -\sum \frac{\partial S}{\partial x_i} dx_i + \sum \frac{\partial u}{\partial x_i} dx_i = \sum \frac{\partial (u - S)}{\partial x_i} dx_i,$$

$u(x) = S(x) + c$, where c is a constant. For $e^{-i\theta}a(x)$ to be constant along this lift, we must have

$$e^{-i(S(x)+c)}a(x) = \text{constant},$$

so $a(x)$ must be a constant multiple of $e^{iS(x)}$. The presence of this multiplicative constant is unavoidable, since the function $S(x)$ is determined by L only up to an additive constant.

If we replace dS by an arbitrary closed 1-form ω on X, the lagrangian submanifold $L = \omega(X)$ is quantizable just when the function $u(x)$ satisfying $du = \omega$ is globally defined modulo 2π; this is the case when the integral of ω around every closed curve in X is an integer multiple of 2π, or, equivalently, when the class $[\omega/2\pi] \in H^1(X; \mathbf{R})$ is in the image of $H^1(X; \mathbf{Z})$. In this case, the function $e^{iu(x)}$ is globally defined and may be associated with $\omega(X)$.

We have now associated, to each quantizable lagrangian section $L = \omega(X) \subseteq T^*X$, a function on X. To get a 1/2-density on X, we should add one piece of information – a 1/2-density on L, which we can pull back to X and multiply by the function. This suggests that the objects to be quantized in general should be pairs consisting of a lagrangian submanifold L and a 1/2-density on L. Such a pair will be called a *quasi-classical state* (see **[SL 2]**). (In case L is the graph of a symplectomorphism $f\colon P_1 \longrightarrow P_2$, it carries a natural 1/2-density induced from the symplectic form on P_1 or P_2.)

Remaining within cotangent bundles, we may ask next how to quantize a quasi-classical state (L, δ) in T^*X for which L does *not* project diffeomorphically onto X. How, for example, should we quantize the fibre $x = 0$ in $T^*\mathbf{R}$? Any function on $T^*X \times S^1$ of the form $\varphi(x, \xi, \theta) = e^{-i\theta}a(x)$ is already constant along the horizontal lifts of the fibre $x = 0$ (as it is along the horizontal lifts of all fibres), so none is distinguished by that condition. Since the fibre $x = 0$ lies only over the origin in $\mathbf{R}$, it is tempting to quantize it by a Dirac delta "function" supported at the origin. It turns out that argument by continuity leads to the same conclusion, as we shall now see.

Consider the line L defined by $x = 0$ in the (x, ξ)-plane, equipped with the 1/2-density $\rho = C|d\xi|^{1/2}$, where C is a constant. It is the limit as $m \longrightarrow 0$ of the line L_m defined by $x = m\xi$, equipped with the 1/2-density ρ_m whose expression in the ξ-coordinate is $C|d\xi|^{1/2}$. Now L_m is also defined by the equation $\xi = x/m$, so it is $dS(\mathbf{R})$, where $S(x) = x^2/2m$; in the x-coordinate, our 1/2-density becomes

$$C\left|d\left(\frac{x}{m}\right)\right|^{1/2} = \frac{C}{|m|^{1/2}}\,|dx|^{1/2}.$$

By our quantization rule for sections of T^*X, we should associate with (L_m, ρ_m) the 1/2-density $(C/|m|^{1/2})e^{ix^2/2m}|dx|^{1/2}$ on X. (This is not in L^2; recall, however, that the objects associated with lagrangian submanifolds of (P, Ω) are generally contained in some *extension* of the Hilbert space which quantizes (P, Ω).) Since $(L, \rho) = \lim_{m\to 0}(L_m, \rho_m)$, it is natural to try to quantize (L, ρ) by

$$\lim_{m\to 0} \frac{C}{|m|^{1/2}}\, e^{ix^2/2m}\,|dx|^{1/2}. \tag{10.1}$$

We cannot make sense out of the last limit in the space of C^∞ 1/2-densities on $\mathbf{R}$, but the limit does exist in the following "weak" sense. If $v = v(x)|dx|^{1/2}$ is any C^∞ 1/2-density on $\mathbf{R}$ with compact support and $u = u(x)|dx|^{1/2}$ is any C^∞ 1/2- density on $\mathbf{R}$, we can form their product by

$$\langle u, v\rangle = \int_{\mathbf{R}} u(x)\bar{v}(x)\, dx.$$

In this way, the space $E(|\mathbf{R}|^{1/2})$ of C^∞ 1/2-densities on $\mathbf{R}$ is identified with a space of linear functionals on the space $\mathcal{D}(|\mathbf{R}|^{1/2})$ of compactly supported C^∞ 1/2-densities. The full space of linear functionals on $\mathcal{D}(|\mathbf{R}|^{1/2})$, continuous with respect to a certain C^∞ topology (see [S]), is denoted by $\mathcal{D}'(|\mathbf{R}|^{1/2})$. Since $\mathcal{D}(|\mathbf{R}|^{1/2})$ contains $E(|\mathbf{R}|^{1/2})$ (as a dense subset), its elements are called *generalized* 1/2*-densities*, or 1/2*-density-valued distributions*, on $\mathbf{R}$.

Now we may try to find the limit (10.1) in $\mathcal{D}'(|\mathbf{R}|^{1/2})$. To do this, we must evaluate

$$\lim_{m\to 0} \int \frac{C}{|m|^{1/2}}\, e^{ix^2/2m}\, \bar{v}(x)\, dx \tag{10.2}$$

where $v(x)$ is a compactly supported C^∞ function on $\mathbf{R}$. In fact, the *principle of stationary phase* tells us that the limit (10.2) exists if m approaches 0 with a fixed sign; it is equal to

$$C(2\pi)^{-1/2}\, e^{(i\pi/4)\,\mathrm{sgn}\, m}\, \bar{v}(0).$$

The functional $v(x)|dx|^{1/2} \longmapsto \bar{v}(0)$ belongs to $\mathcal{D}'(|\mathbf{R}|^{1/2})$; it is called a Dirac delta functional at the origin and will be denoted by $\delta_0 |dx|^{1/2}$. Then we have, in $\mathcal{D}'(|\mathbf{R}|^{1/2})$,

$$\lim_{m\to\pm 0} \frac{C}{|m|^{1/2}}\, e^{ix^2/2m}\, |dx|^{1/2} = C(2\pi)^{-1/2}\, e^{\pm i\pi/4}\, \delta_0\, |dx|^{1/2}.$$

Thus, we are lead to quantize the quasi-classical state $(x = 0, C\,|d\xi|^{1/2})$ by the distribution (determined up to a factor of i) $C(2\pi)^{-1/2} e^{\pm i\pi/4}\, \delta_0(|dx|)^{1/2}$ on X.

We shall now describe a general construction which is motivated by this example. Let V be a real vector space of dimension n. Any lagrangian subspace L of $T^*V = V \times V^*$ which is transversal to the "vertical" space $\{0\} \times V^*$ (which we will denote simply by V^*) is the graph of a symmetric mapping $A_L\colon V \longrightarrow V^*$; equivalently, $L = dS_L(V)$, where S_L is the quadratic function $S_L(x) = \tfrac{1}{2} A_L(x)(x)$. (We remove the indeterminacy in S_L by requiring $S_L(0) = 0$.) If ρ is any 1/2-density on L, we may regard it as well as a 1/2-density on V by pullback, and we quantize the pair (L, ρ) by the 1/2-density $e^{iS_L(x)}\rho$ on V.

If L is the vertical space, its quantization should be a delta functional supported at $0 \in V$. Again, the principle of stationary phase justifies this choice, but instead of working out the details of this we will pass immediately to the general problem of quantizing a pair (L, ρ), where L is an arbitrary lagrangian subspace of T^*V, and ρ is a translation-invariant 1/2-density on L.

The lagrangian subspace L projects onto a subspace $W_L \subseteq V$. We may guess that the quantization of (L, ρ) should consist of distributions which are supported on W_L, but which distributions should they be? The subspace $F = W_L \oplus V^*$ is coisotropic in $V \oplus V^*$, and $F^\perp = 0 \oplus W_L^\perp$, where $W_L^\perp \subseteq V^*$ is the usual annihilator of W_L. Since $V^*/W_L \approx W_L^*$, the reduced symplectic space $F/F^\perp$ is naturally isomorphic to $W_L \oplus W_L^*$. The reduction $(L \cap F)/(L \cap F^\perp)$ of L to $F/F^\perp \approx W_L \oplus W_L^*$ projects onto W_L, so it is the graph of a symmetric mapping $W_L \xrightarrow{A_L} W_L^*$.

This suggests that we quantize (L, ρ) by the function $e^{(1/2)iA_L(x)(x)}$ on W_L times some "delta functional" along W_L. In fact, some juggling of 1/2-densities and exact sequences (see [GU] or [W 7]) shows that there is a natural isomorphism j_L between the space $|L|^{1/2}$ of 1/2-densities on the vector space L and the space $\mathrm{Hom}(|V|^{1/2}, |W_L|)$ of homomorphisms from 1/2-densities on V to 1-densities on W_L. (In case $W_L = V$, j_L is just the pull back isomorphism.) The latter space may be identified with a subspace of $\mathcal{D}'(|V|^{1/2})$: Given $\sigma\colon |V|^{1/2} \longrightarrow |W_L|$ and a "test-density" $u \in \mathcal{D}(|V|^{1/2})$, we may apply σ to it and restrict to W_L, obtaining an element of $\mathcal{D}(|W_L|)$ which may be integrated over W_L to give a complex number. Thus we are led to quantize (L, ρ) by the functional

$$u \longmapsto \int_{W_L} e^{(1/2)iA_L(x)(x)}[(j_L\rho)\circ u],$$

which we will denote by $\delta(L, \rho)$.

The 1-dimensional case suggests that we should modify $\delta(L, \rho)$ by constant factors of $(2\pi)^{1/2}$ and $e^{i\pi/4}$. We shall now show that, with such modifications, the quantization of lagrangian subspaces becomes a continuous process.

We may study the process $(L, \rho) \longrightarrow \delta(L, \rho)$ and, incidentally, define a differentiable structure on the lagrangian grassmannian $L(T^*V)$, by using a special covering of $L(T^*V)$. For each $K \in L(T^*V)$ which is transversal to $\{0\} \times V^*$ (which we denote simply by V^*), we define U_K to be the set of $L \in L(T^*V)$ which are transversal to K. Given any $L \in L(T^*V)$, there is a K which is transversal to both L and V^*, so the U_K's cover $L(T^*V)$.

Now we may identify U_K with the space of quadratic forms on V^*. In fact, the lagrangian splitting $T^*V = V^* \oplus K$ induces an isomorphism of K with V^{**} (it is the negative of the isomorphism of K with V given by projection along V^*) as in Lecture 2, and hence an isomorphism of T^*V with $V^* \oplus V^{**} = T^*(V^*)$. Now each L in U_K is identified with a lagrangian subspace $\alpha_K(L)$ of T^*V^* which is transverse to V^{**}. By the earlier construction in this section, we then have the symmetric mapping $V^* \xrightarrow{A_{\alpha_K(L)}} V^{**}$ and the quadratic function $S_{\alpha_K(L)}(\xi) = \frac{1}{2}A_{\alpha_K(L)}(\xi)(\xi)$ on V^*. We will write $A(V^*|L|K)$ for $A_{\alpha_K(L)}$.

The mappings $U_K \xrightarrow{\varphi_K} \mathrm{Sym}(V^*, V^{**})$ defined by $\varphi_K(L) = A(V^*|L|K)$, which are bijective, will be taken as the charts for $L(T^*V)$. To show that the charts give a differentiable structure, we must study the transition map on $U_{K_1} \cap U_{K_2}$. Writing $A(V|K|V^*)$ for the symmetric map from V to V^* of which K is the graph, and identifying V with V^{**} in the usual way, we have the following lemma, whose proof we omit.

LEMMA. (a) *If $L \in U_{K_2}$, then L lies in U_{K_1} if and only if the operator* $I - A(V^*|L|K_2)[A(V|K_2|V^*) - A(V|K_1|V^*)]$ *is invertible.*

(b) *If $L \in U_{K_1} \cap U_{K_2}$, then*

$$\varphi_{K_1}(L) = \{I - \varphi_{K_2}(L)[A(V|K_2|V^*) - A(V|K_1|V^*)]\}^{-1}\varphi_{K_2}(L).$$

This lemma implies immediately that the charts (U_K, φ_K) define a differentiable structure on $L(T^*V)$.

We return now to the quantization of lagrangian subspaces. The correspondence

$(L, \rho) \longmapsto \delta(L, \rho)$ is clearly continuous for those L which are transversal to the vertical. We look next at those L which are transversal to the "horizontal", i.e. $L \in \mathcal{U}_V$ where we write V for $V \oplus \{0\}$. The trick is to reduce this to the first case by using the Fourier transform, which is the analytic analogue of interchanging V and V^*.

Suppose for the moment that L is transversal to both the horizontal and the vertical. It will help to use coordinates in what follows, so let $(x_1, \ldots, x_n)$ be coordinates on V and $(\xi_1, \ldots, \xi_n)$ the dual coordinates on V^*. We write $|dx|^{1/2}$ for $|dx_1 \wedge \cdots \wedge dx_n|^{1/2}$ and the same for the ξ's. L is described by the equation $\xi = Ax$, where $A = A(V|L|V^*)$ is symmetric and invertible (since L is transversal to the horizontal) and $\rho = c|dx|^{1/2}$ for some constant c. $\delta(L, \rho)$ is then $ce^{(1/2)iA(x)(x)}|dx|^{1/2}$, and its Fourier transform is

$$F\delta(L, \rho) = \left\{\int e^{-i\langle \xi, x\rangle} ce^{(i/2)A(x)(x)}\, dx\right\} |d\xi|^{1/2}.$$

(The "natural" Fourier transform on 1/2-densities involves the canonical 1/2-density $|dx|^{1/2}|d\xi|^{1/2}$ on T^*V.) This Fourier transform is computed, for instance, on pp. 144–145 of [HÖ]. It is

$$F\delta(L, \rho) = c(2\pi)^{n/2} e^{i(\pi/4)\operatorname{sgn} A} e^{-(i/2)A^{-1}(\xi)(\xi)} |\det A|^{-1/2} |d\xi|^{1/2},$$

where sgn A is the signature of the symmetric form A. Now we may interpret the formula for $F\delta(L, \rho)$ in terms of the coordinate φ_V on $\mathcal{U}_V$. In fact, $B = A(V^*|L|V)$ is the negative of the inverse of $A = A(V|L|V^*)$, and $|\det A|^{-1/2}|d\xi|^{1/2} = |dx|^{1/2}$ on L, so $c|\det A|^{-1/2}|d\xi|^{1/2}$ is just the pullback $\bar{\rho}$ of ρ to V^* by the projection along V, so we may write

$$F\delta(L, \rho) = (2\pi)^{n/2} e^{-(i\pi/4)\operatorname{sgn} B} e^{(i/2)B(\xi)(\xi)} \bar{\rho}. \tag{10.3}$$

If L is not transversal to the vertical, we get a similar result. For instance, if $L = V^*$ and $\rho = c|d\xi|^{1/2}$, then $\delta(V^*, \rho) = c\delta_0 |dx|^{1/2}$, and

$$F\delta(V^*, \rho) = c|d\xi|^{1/2} = \bar{\rho},$$

which we can write in the form (10.3) if we remove the factor $(2\pi)^{n/2}$, since $B = A(V^*|V^*|V)$ is zero in this case.

A similar computation shows that, for general $L \in \mathcal{U}_V$,

$$F\delta(L, \rho) = (2\pi)^{(1/2)\dim W_L} e^{-(i\pi/4)\operatorname{sgn} B} e^{(i/2)B(\xi)(\xi)}\, \bar{\rho} \tag{10.4}$$

where $B = A(V^*|L|V)$ and $\bar{\rho}$ is the pullback of ρ to V^*.

We can see now how to "correct" $\delta(L, \rho)$. The signature of B is congruent mod 2 to the rank of B, which is in turn equal to the dimension of W_L. It follows from (10.4) that, except for "multiplicative jumps" of powers of $ei\pi/2 = i$ when the signature of B changes, the map

$$(L, \rho) \longmapsto (2\pi)^{-(1/2)\dim W_L} e^{i(\pi/4)\dim W_L} \delta(L, \rho)$$

is continuous for $L \in \mathcal{U}_V$. (The jumps occur when L fails to be transversal to V^*.)

To take into account the jumps, we may associate to (L, ρ) the 4-tuple

$$e(L, \rho) = \left\{(2\pi)^{-(1/2)\dim W_L} e^{i(\pi/4)\dim W_L} i^k \delta(L, \rho) \,|\, k = 0, 1, 2, 3\right\}$$

of distributions, which *does* depend continuously on (L, ρ).

Finally, we must see what happens on U_K for K other than V. It turns out that the map $(L, \rho) \longmapsto e(L, \rho)$ is still continuous – instead of just the Fourier transform, one must consider

$$F(e^{-(i/2)A(V^*|L|K)^{-1}(x)(x)}\delta(L, \rho)).$$

Now consider the set $M(T^*V) = \{(L, \delta(L, \rho)) \mid \rho \in |L|^{1/2}\} \subseteq L(T^*V) \times \mathcal{D}'(|V|^{1/2})$, with the projection $M(T^*V) \longrightarrow L(T^*V)$ given by $(L, \delta(L, \rho)) \longmapsto L$. Our calculations show that $M(T^*V)$ is a complex line bundle over $L(T^*V)$ whose structure group reduces to the discrete group of multiplications by $\{1, i, -1, -i\}$. Analysis of the jumps when L is not transversal to V shows that this bundle is exactly the Maslov line bundle used in the theory of Fourier integral operators. We have seen, therefore, an "analytic" realization of the Maslov bundle arising from quantization theory.

What structures on T^*V did we use to construct the Maslov bundle? Besides the symplectic structure and linear structure, we used the polarization given by the vertical space V^*, but the horizontal space V played no essential role. In general, if we have a symplectic vector space E with a distinguished linear real polarization π, we may prequantize E as a symplectic manifold and consider the distribution space $\mathcal{D}'(E, \pi)$ in which the Hilbert space obtained by geometric quantization sits as a dense subspace. (The choice of a horizontal space enables one to identify $\mathcal{D}'(E, \pi)$ with $\mathcal{D}'(|E/\pi|^{1/2})$.) In the bundle $L(E) \times \mathcal{D}'(E, \pi)$ over $L(E)$, there is then a distinguished "Maslov bundle" $M(E, \pi)$. This construction can be applied fibre by fibre, if we have a symplectic vector bundle $E \longrightarrow \Omega$ with two real polarizations to produce a line bundle over B. If $E = T_L(T^*X)$, where $L \subseteq T^*X$ is a lagrangian submanifold and the polarizations are given by TL and T_L (fibres), one recovers the Maslov bundle over L used in **[HÖ]**. Its holonomy is the mod 4 reduction of the Maslov class which we defined in Lecture 6.

This analytic construction of the Maslov bundle puts the theory of Fourier integral operators in a new perspective and suggests an extension of Hörmander's symbol construction to arbitrary distributions (see **[WE 6]** and **[WE 7]**). On the other hand, the theory of Fourier integral operators itself provides a means for quantizing certain lagrangian *submanifolds.* We refer the reader to **[D]** and **[HÖ]** for expositions of this theory, mentioning only that the phase functions of Lecture 6 play an important role.

References[1]

[AB] R. Abraham and J. E. Marsden, *Foundations of mechanics,* 2nd edition, Benjamin/Cummings, Reading, 1978.

[AK] Y. Akyildiz, *Dynamical symmetries of the Kepler problem,* Thesis, Univ. of California at Berkeley, 1976.

[AN 1] V. I. Arnol'd, *Sur une propriété topologique des applications globalement canoniques de la mécanique classique,* C. R. Acad. Sci. Paris **261** (1965), 3719–3722. MR **33** #1861.

[AN 2] ———, *Characteristic class entering in conditions of quantization,* Funkcional Anal. i Priložen **1** (1967), 1–14 = Functional Anal. Appl. **1** (1967), 1–13. MR **35** #2296.

[AN 3] ———, *Commentary on Poincaré's geometric theorem,* Selected Works of Henri Poincaré, Vol. II: New methods in celestial mechanics. Topology. Number theory, "Nauka", Moscow, 1972. (Russian) MR **52** #5337.

[AN 4] ———, *Mathematical methods of classical mechanics,* Graduate Texts in Math., No. 60, Springer-Verlag, New York, 1978.

[AN 5] ———, *Critical points of smooth functions,* Proc. Internat. Congr. of Mathematicians (Vancouver, 1974), Vol. 1, Canad. Math. Congress, 1975, pp. 19–39.

[AN V] V. I. Arnol'd and A. Avez, *Ergodic problems of classical mechanics,* Gauthier-Villars, Paris, 1967; English transl., Benjamin, New York and Amsterdam, 1968. MR **35** #334; **38** #1233.

[AT] E. Artin, *Geometric algebra,* Interscience, New York, 1957. MR **18**, 553.

[AU] L. Auslander and B. Kostant, *Polarization and unitary representations of solvable Lie groups,* Invent. Math. **14** (1971), 255–354. MR **45** #2092.

[BA] R. Barrar, *Periodic orbits of the second kind,* Indiana Univ. Math. J. **22** (1972), 33–41.

[BE] F. A. Berezin, *Quantization,* Izv. Akad. Nauk SSSR Ser. Mat. **38** (1974), 1116–1175 = Math. USSR Izv. **8** (1974), 1109–1168.

[BEN] P. Bernat et al., *Representations des groupes de Lie resolubles,* Dunod, Paris, 1972.

[BL 1] R. J. Blattner, *Quantization and representation theory,* Proc. Sympos. Pure Math., vol. 26, Amer. Math. Soc., Providence, R.I., 1973, pp. 147–165. MR **49** #6277.

[BL 2] ———, *The metalinear geometry of non-real polarizations,* Proc. Conf. on Differential Geometrical Methods in Mathematical Physics, Bonn, 1975 (to appear).

[1] References were updated for the 1979 printing.

[BO] R. Bott, *On the iteration of closed geodesics and the Sturm intersection theory,* Comm. Pure Appl. Math. **9** (1956), 176–206. MR **19**, 859.

[C] P. Chernoff and J. Marsden, *Properties of infinite dimensional Hamiltonian systems,* Lecture Notes in Math., vol. 425, Springer-Verlag, Berlin and New York, 1974.

[D] J. J. Duistermaat, *Fourier integral operators,* New York Univ., 1973.

[F] F. T. Farell, *The obstruction to fibering a manifold over a circle,* Bull. Amer. Math. Soc. **73** (1967), 737–740. MR **35** #6151.

[GO] H. Goldstein, *Classical mechanics,* Addison-Wesley, Reading, Mass., 1951. MR **13**, 291.

[GR] R. Greene and K. Shiohama,Trans. Amer. Math. Soc. (to appear).

[GRM] M. L. Gromov, *A topological technique for the construction of solutions of differential equations and inequalities,* Internat. Congr. of Mathematicians (Nice, 1970), Vol. 2, Gauthier-Villars, Paris, 1971, pp. 221–225.

[GU] V. W. Guillemin and S. Sternberg, *Geometric asymptotics,* Amer. Math. Soc. (1976).

[H] R. Hermann, *Vector bundles in mathematical physics,* Vol. II, Benjamin, New York, 1970.

[HC] M. W. Hirsch, *Immersions of manifolds,* Trans. Amer. Math. Soc. **93** (1959), 242–276. MR **22** #9980.

[HI] F. Hirzebruch, *Topological methods in algebraic geometry,* 3rd enlarged ed. (transl. by R. L. E. Schwarzenberger), Springer-Verlag, Berlin and New York, 1966. MR **34** #2573.

[HÖ] L. Hörmander, *Fourier integral operators.* I, Acta Math. **127** (1971), 79–183.

[HV] L. Van Hove, *Sur le problème des relations entre les transformations unitaires de la mécanique quantique et les transformations canoniques de la mécaniques classique,* Acad. Roy. Belgique. Bull. Cl. Sci. (5) **37** (1951), 610–620. MR **13**, 519.

[J] N. Jacobson, *Lie algebras,* Interscience, New York and London, 1962. MR **26** #1345.

[KI 1] A. A. Kirillov, *Unitary representations of nilpotent Lie groups,* Uspehi Mat. Nauk **17** (1962), no. 4 (106), 57–110 = Russian Math. Surveys **17** (1962), no. 4, 53–104. MR 25 #5396.

[KI 2] ———, *Elements of representation theory,* "Nauka", Moscow, 1972. (Russian)

[KO] B. Kostant, *Quantization and unitary representations.* I: *Prequantization,* Lectures in Modern Analysis and Applications. III (E. T. Taam, editor), Lecture Notes in Math., vol. 170, Springer-Verlag, Berlin and New York, 1970, pp. 87–208. MR **45** #3638.

[LA 1] J. L. Lagrange, *Mémoire sur la théorie des variations des éléments des planètes,* Mém. Cl. Sci. Math. Phys. Inst. France (1808), 1–72.

[LA 2] ———, *Second mémoire sur la théorie de la variation des constantes arbitraires dans les problèmes de mécanique,* Mém. Cl. Sci. Math. Phys. Inst. France (1809), 343–352.

[LB] F. Laudenbach, *Formes differentielles de degre* 1 *fermées non singulières: Classes d'homotopie de leurs noyaux* (to appear).

[LN] H. B. Lawson, Jr., *Lectures on the quantitative theory of foliations,* CBMS Regional Conf. Ser. in Math., Amer. Math. Soc., Providence, R.I., No. 27, 1977.

[LS] J. A. Lees, *On the classification of lagrange immersions,* Duke Math. J. **43** (1976), 217–224.

[MA] G. W. Mackey, *The mathematical foundations of Quantum Mechanics: A lecture-note volume,* Benjamin, New York and Amsterdam, 1963. MR **27** #5501.

[MD] J. Marsden, *Darboux's theorem fails for weak symplectic structures,* Proc. Amer. Math. Soc. **32** (1972), 590–592. MR **45** #2755.

[ML] V. P. Maslov, *Theory of perturbations and asymptotic methods,* Moscow State Univ., Moscow, 1965 (Russian); French transl.: *Théorie des perturbations et methodes asymptotiques,* Dunod, Gauthier-Villars, Paris, 1972.

[MO] J. Moser, *On the volume elements on a manifold,* Trans. Amer. Math. Soc. **120** (1965), 286–294. MR **32** #409.

[N] S. P. Novikov, *Algebraic construction and properties of Hermitian analogs of K-theory over rings with involution from the viewpoint of Hamiltonian formalism. Applications to differential topology and the theory of characteristic classes.* I, II, Izv. Akad. Nauk SSSR Ser. Mat. **34** (1970), 253–288; ibid. **34** (1970), 475–500 = Math. USSR-Izv. **4** (1970), 257–292; ibid. **4** (1970), 479–505. MR **45** #1994.

[O] E. Onofri, *Dynamical quantization of the Kepler manifold,* J. Mathematical Phys. **17** (1976), 401–408.

[R] L. P. Rothschild and J. A. Wolf, *Representations of semisimple groups associated to nilpotent orbits,* Ann. Sci. École Norm. Sup. (4) **7** (1974), 155–174 (1975). MR **50** #10158.

[S] L. Schwartz, *Théorie des distributions,* Hermann, Paris, 1966. MR **35** #730.

[SE] I. E. Segal, *Quantization of nonlinear systems,* J. Mathematical Phys. **1** (1960), 468–488. MR **24** #B1144.

[SEM] M. A. Semenov-Tjan-Šanskiĭ, *On a property of Kirillov's integral,* Differential Geometry, Lie Groups, and Mechanics (L. D. Fadde'ev, editor), Zap. Naučn. Sem. Leningrad. Otdel. Mat. Inst. Steklov (LOMI) **37** (1973), 53–65. (Russian) MR **48** #1840.

[SH] V. I. Šubov, *The decomposition of quasiregular representation of a Lie group with the aid of the orbit method,* Differential Geometry, Lie Groups and Mechanics (L. D. Fadde'ev, editor), Zap. Naučn. Sem. Leningrad. Otdel. Mat. Inst. Steklov (LOMI) **37** (1973), 77–96. (Russian) MR **49** #6278.

[SI] C. L. Siegel and J. Moser, *Lectures on celestial mechanics,* Springer-Verlag, Berlin and New York, 1971.

[SIM] D. J. Simms, *Metalinear structures and a geometric quantisation of the harmonic oscillator,* Colloques Internat. du CNRS **237** (1976), pp. 163–173.

[SL 1] J. J. Sławionowski, *Quantum relations remaining valid on the classical level,* Rep. Mathematical Phys. **2** (1971), no. 1, 11–34. MR **44** #2447.

[SL 2] ———, *Geometry of Van Vleck ensembles,* Rep. Mathematical Phys. **3** (1972), no. 3, 157–172. MR **46** #10341.

[SM] S. Smale, *Topology and mechanics.* I, II, Invent. Math. **10** (1970), 305–331; ibid. **11** (1970), 45–64. MR **46** #8263; **47** #9671.

[SN] J. Śniatycki and W. M. Tulczyjew, *Generating forms of Lagrangian submanifolds,* Indiana Univ. Math. J. **22** (1972/73), 267–275. MR **46** #4427.

[SR] J.-M. Souriau, *Structure des systèmes dynamiques,* Dunod, Paris, 1970. MR **41** #4866.

[ST] N. E. Steenrod, *The topology of fibre bundles,* Princeton Univ. Press, Princeton, N.J., 1951. MR **12**, 522.

[SW] R. C. Swanson, *Linear symplectic structures on Banach spaces,* Univ. of California, Santa Cruz, 1976 (preprint).

[TI] D. Tischler, *On fibering certain foliated manifolds over* S^1, Topology **9** (1970), 153–154. MR **41** #1069.

[W 1] A. Weinstein, *Singularities of families of functions,* Differentialgeometrie im Grossen (W. Klingenberg, editor), Bibliographisches Institut, Mannheim, 1971, pp. 323–330.

[W 2] ———, *Symplectic manifolds and their lagrangian submanifolds,* Advances in Math. **6** (1971), 329–346. MR **44** #3351.

[W 3] ———, *Lagrangian submanifolds and hamiltonian systems,* Ann. of Math. (2) **98** (1973), 377–410.

[W 4] ———, *Normal modes for nonlinear hamiltonian systems,* Invent. Math. **20** (1973), 47–57. MR **48** #6564.

[W 5] ———, *Fourier integral operators, quantization, and the spectra of riemannian manifolds,* Colloques Internat. du CNRS **237** (1976), 289–298.

[W 6] ———, *The principal symbol of a distribution,* Bull. Amer. Math. Soc. **82** (1976), 548–550.

[W 7] ———, *The order and symbol of a distribution,* Trans. Amer. Math. Soc. **241** (1978), 1–54.

[WE] H. Weyl, *The classical groups. Their invariants and representations,* Princeton Univ. Press, Princeton, N.J., 1946. (1st ed., 1939; MR **1**, 42.)

[WH] E. T. Whittaker, *A treatise on the analytical dynamics of particles and rigid bodies,* 4th ed., Dover, New York, 1944. MR **6**, 74.

CDEFGHIJ–AMS–89876543